AUSTRALIAN MATHEMATICAL SOCIETY LECTURE SERIES

Editor-in-Chief: Professor J.H. Loxton, School of Mathematics, Physics, Computing and Electronics, Macquarie University, NSW 2109, Australia

Editors:
Professor C.J. Thompson, Department of Mathematics, University of Melbourne, Parkville, Victoria 3052, Australia
Professor C. C. Heyde Department of Statistics, University of Melbourne, Parkville, Victoria 3052, Australia

Low Rank Representations and Graphs for Sporadic Groups

Cheryl E. Praeger

Department of Mathematics
The University of Western Australia

Leonard H. Soicher

School of Mathematical Sciences
Queen Mary and Westfield College, London

CAMBRIDGE
UNIVERSITY PRESS

MATH
.08195857

PUBLISHED BY THE PRESS SYNDICATE OF THE UNIVERSITY OF CAMBRIDGE
The Pitt Building, Trumpington Street, Cambridge CB2 1RP, United Kingdom

CAMBRIDGE UNIVERSITY PRESS
The Edinburgh Building, Cambridge, CB2 2RU, United Kingdom
40 West 20th Street, New York, NY 10011-4211, USA
10 Stamford Road, Oakleigh, Melbourne 3166, Australia

First published 1997

Printed in the United Kingdom at the University Press, Cambridge

A catalogue record for this book is available from the British Library

Library of Congress Cataloguing in Publication data available

ISBN   0 521 56737 8 paperback

Contents   MATH

Preface

This book is a contribution to the study of the sporadic simple groups, which are the twenty-six fascinating finite simple groups which do not belong to any of the infinite families of finite simple groups. In particular, we classify and study all graphs admitting a sporadic simple group or its automorphism group as a vertex-transitive group of automorphisms of rank at most 5. We do this by first finding all the representations of these sporadic groups as transitive permutation groups of rank at most 5. The classification, construction and analysis of these representations and graphs involve both theoretical arguments about permutation groups and characters, and computational methods involving the use of various computer systems for group theory, character theory and graph theory. The ATLAS of Finite Groups was an invaluable resource throughout all our work.

We have tried to make most of our techniques accessible to a beginning graduate student who is willing to study some basic computational group theory. In particular, the construction and analysis of collapsed adjacency matrices are spelled out in detail. For the theoretical analysis of certain permutation representations, some knowledge of permutation group theory is assumed.

Using the presentations for sporadic groups which are given in this book, and the generating sets for the point stabilizers, the reader should be able to construct and study most of the representations and graphs described in this book. Indeed, one of our main aims is to give inexperienced readers sufficient background to enable them to construct and explore finite vertex-transitive graphs.

It is our pleasure to thank the people and organizations who have helped bring this book to fruition. This book would not have been undertaken without the GAP computer system, and we thank its main architect, Martin Schönert, and the leader of the GAP project, Joachim Neubüser, who is also responsible for the open and challenging atmosphere of Lehrstuhl D für Mathematik at the RWTH Aachen, where both authors have spent fruitful visits. Also, many thanks are due to the computational representation theory team at Lehrstuhl D, who put so much effort into the character theory part of GAP. Here special thanks go to Klaus Lux and Thomas Breuer. Thanks are also due to John Cannon and his team at the University of Sydney who produced the Cayley group theory system (and its successor the algebra system MAGMA), which was very useful in our work. We thank Gerhard Michler and the Institute for Experimental Mathematics, University of Essen, for making computational resources available for the analysis of the degree 9606125 permutation representation of the Lyons group, and Gene Cooperman for providing us with this representation. We also acknowledge supercomputer resources obtained through the UK Engineering and Physical Sciences Research Council and the University of London Computer Centre. Special thanks go to Dima Pasechnik, who computed complete sets of collapsed adjacency matrices for two very large representations, given a single matrix for each. Thanks also go to Peter Cameron, Sasha Ivanov, Derek Holt and Petra Fogarty, for useful discussions and suggestions. Finally, we thank the Australian Research Council and the University of Western Australia for research grants to enable us to work at the same location to initiate the research for this book, and again to complete the writing of the book.

List of Tables

1

Low Rank Permutation Groups

1.1 Introduction

Many interesting finite geometries, graphs and designs admit automorphism groups of low rank. In fact, it was a study of the rank 3 case which led to the discoveries and constructions of some of the sporadic simple groups (see [Gor82]). For several classification problems about graphs or designs, the case where the automorphism group is almost simple is of central importance, and many of the examples have a transitive automorphism group of low rank. This is the case, for example, for the classification problems of finite distance-transitive graphs [BCN89, PSY87], and of finite flag-transitive designs [BDD88, BDDKLS90].

This book presents a complete classification, up to conjugacy of the point stabilizers, of the faithful transitive permutation representations of rank at most 5 of the sporadic simple groups and their automorphism groups. These results, summarized in Chapter 5, filled a major gap in the existing classification results for finite, low rank, transitive permutation groups. For each representation classified, we also give the collapsed adjacency matrices (defined in Section 2.3) for all the associated orbital digraphs. We use these collapsed adjacency matrices to classify the vertex-transitive, distance-regular graphs for these low rank representations, and discover some new distance-regular graphs of diameter 2 (but of rank greater than 3) for the O'Nan group $O'N$, the Conway group Co_2, and the Fischer group Fi_{22}. We also classify the graphs of diameter at most 4 on which a sporadic simple group or its automorphism group acts distance-transitively. It turns out that all these graphs are well-known.

We have tried to give enough information so that the interested reader can duplicate most of our results, and study further the fascinating sporadic groups. In particular, we give presentations for most of the sporadic groups G having permutation representations of rank at most 5, together with sets of words generating the appropriate point stabilizers in G. This information allows the reader with access to a good coset enumeration program (such as those within MAGMA [CP95] and GAP [Sch95]) to reconstruct most of the representations studied in this book.

In the 1970s, R.T. Curtis determined many collapsed adjacency matrices for inclusion in the original Cambridge ATLAS, but these do not appear in the published ATLAS [CCNPW85]. In the early to mid 1980s, the primitive permutation representations of the nonabelian simple groups of order up to 10^6 (excluding the family $L_2(q)$) were analysed in detail from the point of view of cellular rings (or coherent configurations) by A.A. Ivanov, M.H. Klin and I.A. Faradžev [IKF82, IKF84] (see also [FIK90, FKM94]). As part of this analysis, these representations were explicitly constructed using the CoCo computer package [FK91], and all collapsed adjacency matrices for the orbital digraphs were determined. Furthermore, collapsed adjacency matrices have been constructed by others for certain specific orbital digraphs for sporadic groups (see [ILLSS95] and its references), but we have computed all the collapsed adjacency matrices in this book from scratch, using the methods we describe, except for two representations where explicit references are given.

Any permutation representation of rank at most 5 is multiplicity-free (that is, the sum of distinct complex irreducible representations), and for primitive permutation representations, the classification in this book has recently been extended in [ILLSS95], to give a complete classification of the primitive multiplicity-free permutation representations of the sporadic simple groups and their automorphism groups, together with a classification of the graphs Γ on which such a group acts primitively and distance-transitively. It is shown that for such a distance-transitive graph Γ, we have diam(Γ) $\leq$ 4, and so Γ appears in our classification. Even more recently, Breuer and Lux [BL96] have completed the classification of the imprimitive multiplicity-free permutation representations of the sporadic simple groups and their automorphism groups.

1.2 Transitive permutation groups, orbitals and ranks

The *symmetric group* on a set Ω is the group Sym(Ω) of all permutations of Ω. If Ω is finite of cardinality n, then Sym(Ω) is often denoted S_n. A *permutation group* G on a set Ω is a subgroup of Sym(Ω), and G is said to be *transitive* on Ω if, for all $\alpha, \beta \in \Omega$, there is an element $g \in G$ such that the image α^g of α under g is equal to β. More generally, the *orbit* of G containing a point $\alpha \in \Omega$ is the set $\alpha^G := \{\alpha^g \mid g \in G\}$.

For the remainder of the section, let G be a transitive permutation group on a finite set Ω.

The permutation group G on Ω can also be regarded as a permutation group on $\Omega \times \Omega$ by defining

$$(\alpha, \beta)^g = (\alpha^g, \beta^g) \qquad (\alpha, \beta \in \Omega, g \in G).$$

The number of orbits of G on $\Omega \times \Omega$ is called the *rank* of G on Ω.

If α, β are distinct points of Ω, then the pairs (α, α) and (α, β) lie in different orbits of G on $\Omega \times \Omega$. Thus, for $|\Omega| > 1$, the rank of G is at least 2. A permutation group on Ω is said to be *2-transitive* (or *doubly transitive*) on Ω if it is transitive on the ordered pairs of distinct points of Ω. Thus, for $|\Omega| > 1$, the 2-transitive groups are precisely the permutation groups of rank 2. The classification of the finite 2-transitive groups was one of the first consequences for permutation groups of the finite simple group classification, and the problem of classifying finite permutation groups of low rank is a natural extension of this classification.

The orbits of G on $\Omega \times \Omega$ are called *orbitals*, and to each orbital E we associate the directed graph with vertex set Ω and edge set E, the so-called *orbital digraph* for E. It is easy to show that the orbitals for G are in one-to-one correspondence with the orbits on Ω of the *stabilizer* $G_\alpha := \{g \in G \mid \alpha^g = \alpha\}$ of a point $\alpha \in \Omega$. This correspondence maps an orbital E to the set of points $\{\beta \mid (\alpha, \beta) \in E\}$. The orbits of G_α on Ω are called *suborbits* of G, and their lengths are called the *subdegrees* of G.

If G has rank r, then a point stabilizer will have exactly r orbits on Ω, and we say that such a stabilizer is a *rank r subgroup* of G.

1.3 Permutation representations

Let G be a group and Ω a set. An *action* of G on Ω is a function which associates to every $\alpha \in \Omega$ and $g \in G$ an element α^g of Ω such that, for all $\alpha \in \Omega$ and all $g, h \in G$, $\alpha^1 = \alpha$ and $(\alpha^g)^h = \alpha^{gh}$. In a natural way, an action defines a *permutation representation* of G on Ω, which is a homomorphism φ from G into $\mathrm{Sym}\,(\Omega)$: simply define $(g)\varphi \in \mathrm{Sym}\,(\Omega)$ by $\alpha^{(g)\varphi} := \alpha^g$. Conversely, a permutation representation naturally defines an action of G on Ω, leading to a natural bijection between the actions of G on Ω and the permutation representations of G on Ω (see [NST94, pp. 30–32]). Note also that a permutation group H on Ω defines a natural representation (and action) of H on Ω, by defining the representation to be the identity map.

Most of the definitions of Section 1.2 apply to permutation representations by applying them to the permutation group which is the image of that representation. Thus, a permutation representation is said to be transitive if its image is transitive. Similarly, the orbits of a representation are those of its image and, if the representation is transitive, then its rank, orbitals, suborbits and subdegrees are those of its image. However, the point stabilizer $G_\alpha := \{g \in G \mid \alpha^g = \alpha\}$ for the representation may be a proper preimage of the point stabilizer for the permutation group image.

A permutation representation is said to be *faithful* if its kernel is the trivial group of order 1, in which case G is isomorphic to its permutation group image, and we are back to the case of permutation groups. In this book we study faithful representations of the sporadic simple groups and their automorphism groups. If a representation of a (sporadic) simple group is not faithful then clearly its image is the trivial group, and a non-faithful representation of the automorphism group of a sporadic simple group has an image of order 1 or 2 (as a sporadic simple group has index at most 2 in its automorphism group).

1.4 Permutational equivalence and permutational isomorphism

There are several slightly different concepts of equivalence, or isomorphism, for permutation representations and permutation groups (see [NST94, pp. 32–33]). Since an abstract group may be represented in many different ways as a permutation group, the notion of group isomorphism does not provide a sufficiently refined measure for distinguishing between different permutation representations and different permutation groups. The most general concept of permutational equivalence concerns different groups acting on different sets. A *permutational equivalence* of permutation representations of groups G, G^* acting on Ω, Ω^* respectively is a pair (θ, ϕ) of functions, where $\theta : \Omega \to \Omega^*$ is a bijection and $\phi : G \to G^*$ is an isomorphism, and

$$(\alpha^g)\theta = (\alpha\theta)^{g\phi}$$

for all $\alpha \in \Omega$ and all $g \in G$, and the representations (and actions) of G and G^* are said to be (permutationally) equivalent. Clearly θ induces a bijection between the sets of orbits of G, G^* in Ω, Ω^* respectively. Also, the restriction of ϕ to the stabilizer G_α of a point $\alpha \in \Omega$ is an isomorphism onto the stabilizer in G^* of the point $\alpha\theta \in \Omega^*$. Thus the equivalence induces bijections of the sets of orbits and point stabilizers of the two permutation representations. In the particular case of transitive representations of G, G^* on finite sets Ω, Ω^*, the permutational equivalence (θ, ϕ) preserves rank and subdegrees. Moreover, this equivalence induces a second equivalence $(\theta \times \theta, \phi)$ of the natural representations of G, G^* acting on $\Omega \times \Omega$ and $\Omega^* \times \Omega^*$ respectively (namely, by defining $(\alpha, \beta)(\theta \times \theta) := (\alpha\theta, \beta\theta)$ for all $(\alpha, \beta) \in \Omega \times \Omega$), such that $\theta \times \theta$ induces a bijection from the set of orbitals of G in $\Omega \times \Omega$ to the set of orbitals of G^* in $\Omega^* \times \Omega^*$, and preserves the isomorphism classes of the associated orbital digraphs.

If $G = G^*$ then the isomorphism ϕ is an automorphism of G. In the special case where ϕ is the identity map, the equivalence $(\theta, 1)$ is called a *permutational isomorphism*. Thus, roughly speaking, a permutational isomorphism amounts to a relabelling of the point set.

The notions of permutational equivalence and permutational isomorphism for permutation groups G, G^* on Ω, Ω^* respectively, are defined to be the same as these concepts for their natural representations. Note that the classification of faithful permutation representations up to permutational equivalence (respectively isomorphism) is the same as the classification of permutation groups up to permutational equivalence (respectively isomorphism).

In our subsequent discussion we use the following notation: for a group G, Aut G denotes the *automorphism group* of G, Inn G the group of *inner automorphisms* of G, and Out $G :=$ Aut $G/$Inn G is the *outer au-*

tomorphism group of G. Each element of $\operatorname{Aut} G \setminus \operatorname{Inn} G$ is called an *outer automorphism* of G.

Suppose that G has a transitive permutation representation on the set Ω, and choose $\alpha \in \Omega$. Then this representation is permutationally isomorphic to the representation of G, acting by right multiplication, on the right cosets of the point stabilizer G_α [NST94, Theorem 6.3]. If $\varphi \in \operatorname{Aut} G$, then G also has a transitive permutation representation, acting by right multiplication, on the set Ω^* of right cosets of the subgroup $K := (G_\alpha)\varphi$, and

$$\theta : \alpha^g \mapsto K(g\varphi) \quad (\alpha \in \Omega, g \in G)$$

is a well-defined bijection $\theta : \Omega \to \Omega^*$. Moreover the pair (θ, φ) is an equivalence between the permutation representations of G on Ω and on Ω^*. Of course (θ, φ) is by definition a permutational isomorphism if and only if φ is the identity. However the permutation representations of G on Ω and on Ω^* are permutationally isomorphic if and only if G_α and K are conjugate in G [NST94, Theorem 6.3 and Proposition 6.5]. We see from this discussion that, in general, two transitive representations of G are permutationally isomorphic if and only if a point stabilizer for one representation is in the same conjugacy class in G as a point stabilizer for the other representation. Moreover, there is a permutation representation of $\operatorname{Aut} G$ on the set of permutational isomorphism classes of transitive permutation representations of G such that $\operatorname{Inn} G$ is contained in the kernel. So in fact we have a permutation representation induced of $\operatorname{Out} G := \operatorname{Aut} G / \operatorname{Inn} G$ on these permutational isomorphism classes. The orbits of $\operatorname{Aut} G$ (and of $\operatorname{Out} G$) correspond to the permutational equivalence classes of transitive permutation representations of G. Thus the permutational isomorphism classes (respectively permutational equivalence classes) of transitive permutation representations of G are in one-to-one correspondence with the conjugacy classes of subgroups of G (respectively orbits of $\operatorname{Aut} G$, and hence of $\operatorname{Out} G$, on these conjugacy classes).

The classification of transitive permutation representations in this book is up to permutational isomorphism, which is the same as the classification up to conjugacy of the point stabilizers.

Two different permutational isomorphism classes of transitive representations correspond to the same permutational equivalence class if and only if there is an outer automorphism of G mapping one permutational isomorphism class to the other. In the case where G is a sporadic simple group, $|\text{Out}\, G| \leq 2$. Hence, in this situation, an outer automorphism of G will interchange the two permutational isomorphism classes, and will also interchange the corresponding conjugacy classes of point stabilizers. We will point this out whenever it occurs.

1.5 Invariant partitions and primitivity

If G is a permutation group on a set Ω, then a partition P of Ω is said to be *G-invariant* (and G is said to *preserve* P) if the elements of G permute the blocks of P blockwise, that is, for $B \in P$ and $g \in G$, the set B^g is also a block of P. The blocks of a G-invariant partition are called *blocks of imprimitivity* for G. If G is transitive on Ω then all blocks of a G-invariant partition P of Ω have the same cardinality and G acts transitively on P. Moreover, every permutation group G on Ω preserves the two partitions $\{\Omega\}$ and $\{\{\alpha\} \mid \alpha \in \Omega\}$; these are called *trivial partitions* of Ω, and their blocks, Ω and $\{\alpha\}$ for $\alpha \in \Omega$, are called *trivial blocks of imprimitivity*. All other partitions of Ω are said to be *nontrivial*. A permutation group G is said to be *primitive* on Ω if G is transitive on Ω and the only G-invariant partitions of Ω are the trivial ones. Also G is said to be *imprimitive* on Ω if G is transitive on Ω and G preserves some nontrivial partition of Ω.

1.6 The O'Nan-Scott theorem for finite primitive permutation groups

It is not difficult to see that the set of orbits of a normal subgroup of a transitive permutation group G on Ω is a G-invariant partition of Ω. Thus each nontrivial normal subgroup of a primitive permutation group is transitive. In particular, for finite primitive permutation groups G on Ω the *socle* of G, $soc(G)$, which is the product of its minimal normal subgroups, is transitive on Ω. Several different types of finite primitive permutation groups have been identified in the O'Nan-Scott Theorem ([Sco80, AS85] or see [LPS88]) and are described according to the structure and permutation action of their socles.

A finite primitive permutation group G has at most two minimal normal subgroups, and if M, N are distinct minimal normal subgroups of G, then $M \cong N$, M and N are nonabelian, and both act regularly on Ω

(see [Sco80] or [LPS88]). (A permutation group on Ω is *regular* if it is transitive, and only the identity element fixes a point of Ω.)

For most of the types of finite primitive groups, the socle is the unique minimal normal subgroup, and for all types the socle is a direct product of isomorphic simple groups. The types of finite primitive permutation groups are described in [LPS88] as follows. Let G be a primitive permutation group on a finite set Ω, and let $N := soc(G)$. Then $N = T^k$ for some simple group T and positive integer k, and one of the following holds.

Affine type. Here $N = Z_p^k$ (p a prime) is elementary abelian, N is the unique minimal normal subgroup of G, N is regular on Ω, and Ω can be identified with a finite vector space V in such a way that N is the group of translations of V and G is a subgroup of the group $AGL(V)$ of affine transformations of V.

Amost simple type. The socle $N = T$ is a nonabelian simple group ($k = 1$), so $T \leq G \leq \text{Aut}\,T$, that is, G is an *almost simple group*. Also $T_\alpha \neq 1$.

For the remaining types $N = T^k$ with $k \geq 2$ and T a nonabelian simple group.

Simple diagonal type. Here G is a subgroup of the group

$$W := \{(a_1,\ldots,a_k).\pi \mid a_i \in \text{Aut}\,T, \pi \in S_k,$$
$$a_i \equiv a_j \,(\text{mod}\,\text{Inn}\,T)\text{ for all }i,j\},$$

where $\pi \in S_k$ permutes the components a_i naturally. With the obvious multiplication, W is a group with socle $N = T^k$, and $W = N.(\text{Out}\,T \times S_k)$, a (not necessarily split) extension of N by $\text{Out}\,T \times S_k$. The action of W on Ω is equivalent to its action by right multiplication on the set of right cosets of its subgroup

$$W_\alpha := \{(a,\ldots,a).\pi \mid a \in \text{Aut}\,T, \pi \in S_k\} \cong \text{Aut}\,T \times S_k.$$

The group G must contain N, and $N_\alpha = \{(a,\ldots,a) \mid a \in T\}$ is a diagonal subgroup of N, hence the name 'diagonal type'.

Product type. For this type, G is a subgroup of a wreath product $W := H\,\text{wr}\,S_l$ in product action on $\Omega = \Lambda^l$, where $l \geq 2$ and l divides k, H is a primitive permutation group on Λ, $soc(H) \cong T^{k/l}$, and $N = soc(W) = soc(H)^l$ is contained in G. The group H is of either almost simple or simple diagonal type.

Twisted wreath type. For this type, $G = T \operatorname{twr}_\varphi P = T^k.P$ is a twisted wreath product, where $P \leq S_k$, and N is regular on Ω.

More information about the structure of these groups can be found in [AS85, Sco80, LPS88].

1.7 Existing classifications of low rank primitive groups

Long before the description of finite primitive permutation groups that we find in the O'Nan-Scott Theorem had been written down, W. Burnside [Bur11, Section 154] proved that a finite 2-transitive group is of either affine or almost simple type. In fact, the minimum ranks for finite primitive groups of the other types tend to be higher than those for primitive groups of affine or almost simple type, and it follows from the O'Nan-Scott Theorem that the finite primitive groups of rank at most 5 are essentially known once the almost simple ones and the affine ones have been classified (see [Cuy89]). According to the finite simple group classification a nonabelian finite simple group T is either an alternating group, a group of Lie type, or one of the 26 sporadic simple groups (see [Gor82]). Thus the socle T of an almost simple group G is a simple group of one of these types.

The finite 2-transitive groups have been completely classified using the finite simple group classification, and this result is the culmination of the work of many people. The 2-transitive representations of the finite symmetric and alternating groups were classified by E. Maillet [Mai1895] in 1895. Those of the finite almost simple groups of Lie type were determined by C.W. Curtis, W.M. Kantor and G.M. Seitz [CKS76] in 1976, and the classification of the 2-transitive groups of almost simple type was completed and announced by P.J. Cameron [Cam81] in 1981 as a consequence of the finite simple group classification. The finite soluble 2-transitive groups were classified by B. Huppert [Hup57] in 1957; the major part of the classification of the finite insoluble 2-transitive groups of affine type was done by C. Hering [Her74, Her85], and a complete and independent proof of the classification of finite 2-transitive groups of affine type was given by M.W. Liebeck [Lie87, Appendix 1].

A great deal of effort has gone into understanding low rank primitive permutation groups, in particular those of rank at most 5. It follows from the O'Nan-Scott Theorem (see [Cuy89, Corollary 2.2]) that, if G

is primitive of rank at most 5 on a finite set Ω, then either G is of affine
or almost simple type, or G is a subgroup of a wreath product $H \operatorname{wr} S_k$
in product action on $\Omega = \Lambda^k$, where $k \in \{2,3,4\}$ and H is an almost
simple 2-transitive permutation group on Λ, or G has simple diagonal
type with socle isomorphic to $L_2(q) \times L_2(q)$ for some $q \in \{5,7,8,9\}$.
Thus a classification of primitive permutation groups of rank up to 5 is
reduced to a classification of those of affine or almost simple type.

We consider the almost simple case first. In 1972 E.E. Bannai [Ban72]
classified all primitive permutation representations of rank at most 5 of
the finite alternating and symmetric groups. In 1982 W.M. Kantor and
R.A. Liebler [KL82] classified the primitive rank 3 representations of
the classical groups (see also [Sei74]). In 1986 M.W. Liebeck and J. Saxl
[LS86] found all the primitive rank 3 representations of the exceptional
simple groups of Lie type, and A. Brouwer, R.A. Wilson and L.H. Soicher
(see [LS86]) determined those of the sporadic simple groups, thereby
completing the classification of almost simple primitive rank 3 groups.
In 1989, H. Cuypers [Cuy89] completed the classification of all primitive
representations of rank at most 5 of all finite almost simple groups of Lie
type. Part of the purpose of this book is to complete the classification
of the almost simple primitive groups of rank at most 5 by classifying
all such representations of the sporadic almost simple groups. Note that
the sporadic almost simple groups are the sporadic simple groups and
their automorphism groups, since a sporadic simple group has index at
most 2 in its automorphism group.

Finite soluble primitive groups of rank 3 are primitive groups of affine
type and were classified by D.A. Foulser [Fou69] in 1969. The classification of all primitive rank 3 groups of affine type was completed by
M.W. Liebeck [Lie87] in 1987. From these results, and the results presented in this book, it follows that to complete the classification of finite
primitive permutation groups of rank at most 5, only the affine primitive
groups of rank 4 and 5 remain to be classified.

1.8 Low rank sporadic classification

In this book we classify all (primitive and imprimitive) faithful transitive
permutation representations of rank at most 5 of the sporadic simple
groups and their automorphism groups. The rank 2 case is included

for completeness, and the rank 3 case expands the list in [LS86] by classifying all imprimitive rank 3 representations of these groups.

We also provide detailed information about the digraphs on which the permutation groups we describe act vertex-transitively. Background about such graphs is given in Chapter 2, and a discussion of the methods used in our investigations is in Chapter 3. Chapter 4 contains the main body of our work, with the description of the representations and digraphs for the individual sporadic groups, together with many presentations, and Chapter 5 summarizes the representations and distance-regular graphs classified.

2

Digraphs for Transitive Groups

Let G be a transitive permutation group on a finite set Ω. In this chapter we discuss the digraphs with vertex-set Ω on which G acts as a (vertex-transitive) group of automorphisms. We also discuss distance-regularity and distance-transitivity of graphs.

2.1 Some definitions for digraphs

Before proceeding further, it is convenient to record some basic definitions for digraphs.

A *directed graph*, or *digraph*, $\Gamma = (V, E)$ consists of a finite set V, together with a subset E of $V \times V$. The elements of V are called *vertices*, and the elements of E are called *edges*. For $v \in V$, the set

$$\Gamma(v) := \{w \in V \mid (v, w) \in E\}$$

is called the set of *neighbours* of v in Γ.

Let $\Gamma = (V, E)$ be a digraph. A *path* (some would say a directed path) of *length* n in Γ is a sequence $v_1, \ldots, v_{n+1}$ of vertices of Γ such that $(v_i, v_{i+1}) \in E$ for $i = 1, \ldots, n$. We say that such a path *connects* v_1 to v_{n+1}, and that v_1 is *connected* to v_{n+1}. We call Γ *connected* (some would say strongly connected) if v is connected to w for every $v, w \in V$. If v is connected to w in Γ, then we define the *distance* $d(v, w) = d_\Gamma(v, w)$ to be the length of a shortest path connecting v to w. Now if Γ is connected then its *diameter* diam(Γ) is defined to be $\max\{d(v, w) \mid v, w \in V\}$.

The digraph Γ is called *simple* if whenever (v, w) is an edge then (w, v) is an edge and $v \neq w$. If Γ is simple, then we usually consider Γ to be an ordinary undirected graph by identifying each edge (v, w) with (w, v) and speaking of the *undirected edge* $\{v, w\}$. We also call Γ a *simple graph* in this case. For simple graphs the concepts defined above (such as set of neighbours, path, path length, distance, connectivity and diameter) still apply with undirected edges replacing (directed) edges in the definitions.

A *cycle* in a simple (di)graph Γ is defined to be a path of length ≥ 3 connecting a vertex to itself and having no vertices repeated except for the first and last. If a simple graph Γ contains cycles, then we define the *girth* of Γ to be the length of a shortest cycle of Γ; otherwise the girth is defined to be the symbol ∞.

Let $\Gamma = (\Omega, E)$ be a digraph, and G be a permutation group on Ω. We say that G *acts* on Γ if, for all $(\alpha, \beta) \in E$ and $g \in G$, we have $(\alpha^g, \beta^g) \in E$; that is, G is a group of *automorphisms* of Γ. The automorphism group Aut (Γ) of the digraph Γ is the subgroup of Sym (Ω) consisting of all automorphisms of Γ. We say that Γ is *vertex-transitive* if Aut (Γ) is transitive on the vertex-set Ω, and we say that Γ is a *rank r graph* if Aut (Γ) is a transitive group of rank r on Ω.

2.2 Generalized orbital digraphs

Let G be a transitive permutation group on a finite set Ω, and let $E \subseteq \Omega \times \Omega$ be an orbital for G. Recall that the orbital digraph for G associated with E is simply the digraph (Ω, E). Clearly, G is transitive on the vertices and on the edges of an orbital digraph. On the other hand, if $\Gamma = (\Omega, F)$ is a digraph (with vertex-set Ω) on which G acts (as a vertex-transitive group of automorphisms), then F is a union of orbitals for G. Such a digraph Γ is called a *generalized orbital digraph* for G. If in addition, G acts edge-transitively on the generalized orbital digraph $\Gamma = (\Omega, F)$, then F is either empty or a single orbital for G.

One of the orbitals for G is $\{(\alpha, \alpha) \mid \alpha \in \Omega\}$. Its associated orbital digraph consists of a 'loop' (α, α) at each vertex α, and it corresponds to the suborbit $\{\alpha\}$. This orbital, and its corresponding digraph and suborbit are all said to be *trivial*. All others are said to be *nontrivial*. Clearly, each nontrivial orbital digraph has no loops.

Each orbital E has a *paired orbital* E^* defined by

$$E^* := \{(\beta, \gamma) \mid (\gamma, \beta) \in E\}.$$

As E and E^* are orbits for G on $\Omega \times \Omega$, they are either equal or disjoint.

If $E \cap E^* = \emptyset$, then the adjacency relation for the orbital digraph (Ω, E) is anti-symmetric in the sense that if $(\beta, \gamma) \in E$ then $(\gamma, \beta) \notin E$. On the other hand, if $E = E^*$ then, in the orbital digraph (Ω, E), (β, γ) is an edge whenever (γ, β) is. In this case E is said to be *self-paired* and the associated digraph is said to be *undirected*. Moreover, unless the self-paired orbital E is the trivial orbital, the orbital digraph (Ω, E) is simple, in which case it is often called the *orbital graph* associated with E. Similarly, a simple generalized orbital digraph will sometimes be referred to as a *generalized orbital graph*.

It is worthwhile to mention the following result, discovered independently by D.G. Higman [Hig67] and C.C. Sims [Sim67], which is not difficult to prove, but provides a nice characterization of primitivity in terms of orbital digraphs.

Theorem 2.1 *Let G be a transitive permutation group on a finite set Ω. Then G is primitive if and only if each nontrivial orbital digraph for G is connected.*

2.3 Collapsed adjacency matrices

As before, G is a transitive permutation group on a finite set Ω, and $\alpha \in \Omega$. We fix an ordering $\Omega_1 = \{\alpha\}, \Omega_2, \ldots, \Omega_r$ on the distinct orbits of G_α, which have respective representatives $\alpha_1 = \alpha, \alpha_2, \ldots, \alpha_r$.

Let $\Gamma = (\Omega, E)$ be a generalized orbital digraph for G, and for $1 \leq k, j \leq r$, define

$$A[k,j] := |\Gamma(\alpha_k) \cap \Omega_j|.$$

In other words, $A[k,j]$ is the number of neighbours of α_k in Ω_j. It is very important to note that $A[k,j]$ does not depend on the choice α_k of suborbit representative of Ω_k. Indeed, if β is any element of Ω_k, then there is an element $g \in G_\alpha$ such that $\alpha_k^g = \beta$, and we have

$$\Gamma(\beta) \cap \Omega_j = \Gamma(\alpha_k^g) \cap \Omega_j = \Gamma(\alpha_k)^g \cap \Omega_j^g = (\Gamma(\alpha_k) \cap \Omega_j)^g.$$

Thus we have

$$A[k,j] = |\Gamma(\alpha_k) \cap \Omega_j| = |\Gamma(\beta) \cap \Omega_j|.$$

The $r \times r$ integer matrix

$$A = (A[k,j])$$

is called the *collapsed adjacency matrix* for Γ (with respect to G and the suborbit ordering).

For $1 \leq i \leq r$, let A_i be the collapsed adjacency matrix (with respect to G and the suborbit ordering) for the orbital digraph (Ω, E_i) corresponding to the suborbit Ω_i. In particular, A_1 is the $r \times r$ identity matrix. Since $\Gamma = (\Omega, E)$ is a generalized orbital digraph for G, we have that E is the disjoint union of certain orbitals for G, and we have the following obvious, but useful, result.

Proposition 2.2 *As above, let A be the collapsed adjacency matrix for the generalized orbital digraph $\Gamma = (\Omega, E)$, where E is the disjoint union of orbitals, $E_{i_1}, \ldots, E_{i_s}$, for G. Then $A = A_{i_1} + \cdots + A_{i_s}$.* $\square$

In Chapter 4, we shall use the notation $\Gamma\{i_1, \ldots, i_s\}$ for a generalized orbital digraph whose edge set is the disjoint union of orbitals indexed by $i_1, \ldots, i_s$.

The calculation and application of collapsed adjacency matrices is discussed in Chapter 3. Many properties of a generalized orbital digraph Γ can easily be deduced from a collapsed adjacency matrix for Γ. Note that if (θ, ϕ) is a permutational equivalence of the finite transitive permutation groups G, G^*, then G and G^* have exactly the same sequence of collapsed adjacency matrices for their respective sequences of orbital digraphs (with respect to an ordering of the suborbits of G and the image of this ordering under θ).

Remarks In this book, we have chosen to avoid the theory of association schemes (see [BI84] and [BCN89]), and the closely related theory of coherent configurations [Hig75, Hig76]. The theory of coherent configurations was independently developed in the former Soviet Union under the name of cellular rings (see [FIK90]). Readers who wish to advance further in their study of permutation groups and digraphs would be well-advised to study these theories. We do remark, however, that the collapsed adjacency matrices for the orbital digraphs of the finite transitive permutation group G on Ω give the intersection numbers, or structure constants, for the (homogeneous) coherent configuration whose classes are the nontrivial orbitals of G, as follows. For $1 \leq j \leq r$, where r is the rank of G, define j^* by the rule that suborbit Ω_{j^*} corresponds to the paired orbital of the orbital corresponding to Ω_j. Then if E_i is the G-orbital corresponding to Ω_i, the (k, j)-entry $A_i[k, j]$ of the collapsed adjacency matrix A_i for (Ω, E_i) is equal to the intersection number $p^{k-1}_{j-1,i^*-1} = p^{k^*-1}_{i-1,j^*-1}$ as defined in [BCN89, Section 2.1]. This means that, if all the orbitals for G are self-paired, then our collapsed adjacency matrices for the orbital digraphs are the same as the intersection matrices defined in [BCN89, p. 45] for the associated (symmetric) association scheme (except for our indexing starting at 1, rather than 0). Indeed, the collapsed adjacency matrices for the orbital digraphs for G are the transposes of the intersection matrices as originally defined by D.G. Higman in [Hig67] (again, except for our indexing from 1).

2.4 Distance-regularity and distance-transitivity

Suppose now that $\Gamma = (\Omega, E)$ is a connected simple graph, and let $d := \text{diam}(\Gamma)$. For $0 \le i \le d$, define

$$\Gamma_i := \{(\beta, \gamma) \mid d(\beta, \gamma) = i\},$$

the set of pairs of vertices at distance i; and for each vertex β define

$$\Gamma_i(\beta) := \{\gamma \mid (\beta, \gamma) \in \Gamma_i\},$$

the set of vertices at distance i from β. In particular, $\Gamma_1(\beta) = \Gamma(\beta)$. Then Γ is said to be *distance-regular* if there are constants $b_0, \dots, b_{d-1}$, and $c_1, \dots, c_d$, such that, for each $0 \le i \le d$, and each pair β, γ of vertices at distance i, there are precisely c_i neighbours of γ in $\Gamma_{i-1}(\beta)$ (if $i > 0$) and b_i neighbours of γ in $\Gamma_{i+1}(\beta)$ (if $i < d$). The sequence

$$\iota(\Gamma) := \{b_0, b_1, \dots, b_{d-1}; c_1, c_2, \dots, c_d\},$$

is called the *intersection array* of Γ. A distance-regular graph Γ is regular of valency $k := b_0 = |\Gamma(\beta)|$. Set $b_d = c_0 = 0$. The numbers c_i, b_i and a_i, where

$$a_i := k - b_i - c_i$$

is the number of neighbours of γ in $\Gamma_i(\beta)$, for $d(\beta, \gamma) = i$, are called the *intersection numbers* of Γ $(i = 0, \dots, d)$.

If G acts on Γ, then clearly each of the sets Γ_i is fixed setwise by G. Such a group G is said to act *distance-transitively* on Γ if G is transitive on each of the Γ_i, in which case G is transitive on the vertex-set Ω and the orbitals for G are precisely the sets $\Gamma_0, \dots, \Gamma_d$. We say that a simple connected graph Γ is *distance-transitive* if $\text{Aut}(\Gamma)$ acts distance-transitively on Γ, and clearly, if Γ is distance-transitive then it is distance-regular.

The converse is far from true however, as some of our examples in Chapter 4 will illustrate.

If G is a transitive permutation group on Ω, of rank r, and G acts on the simple connected graph $\Gamma = (\Omega, E)$, then we see that G acts distance-transitively on Γ if and only if the diameter of Γ is $r - 1$. Thus, if $r \leq 3$ then each transitive permutation group G of rank r acts distance-transitively on each simple connected orbital digraph for G.

2.5 Graphs for groups of ranks 2 and 3

If G is 2-transitive on a finite set Ω of size $n > 1$ then the subdegrees of G are $1, n - 1$, and there are two orbital digraphs for G, the trivial orbital digraph and the *complete graph* K_n of size n. With respect to G, the complete graph K_n has collapsed adjacency matrix

$$\begin{pmatrix} 0 & n-1 \\ 1 & n-2 \end{pmatrix},$$

and K_n is the unique distance-regular graph with intersection array

$$\{n - 1; 1\}.$$

Now suppose that the transitive permutation group G on Ω has rank 3. Then both nontrivial orbital digraphs for G are simple if and only if one of them is, and this holds if and only if $|G|$ is even (some pair of unequal points in Ω are swapped by some element of G if and only if G contains an involution). By the celebrated Odd Order Theorem of W. Feit and J.G. Thompson [FT63], if G is insoluble, as in the almost simple case, then G has even order. If G is primitive of rank 3 then each of its nontrivial orbital digraphs is connected. Thus, if G is a primitive rank 3 group of even order then its nontrivial orbital digraphs form a pair of complementary connected simple graphs on each of which G acts distance-transitively. (The *complement* of a simple graph (V, E), where E is a set of undirected edges, is $(V, \bar{E})$, where $\bar{E}$ is the set of 2-subsets of V not in E.)

If G is an imprimitive rank 3 group, having a blocks of imprimitivity each of size b, then G has subdegrees $1, b - 1, b(a - 1)$. It is easy to see that both nontrivial orbital digraphs for G are simple, one being $a.K_b$, a disjoint copies of K_b, and the other being its complement, the *complete multipartite graph* $K_{a \times b}$, on which G acts distance-transitively.

The graph $K_{a \times b}$ is the unique distance-regular graph with intersection array

$$\{b(a-1), b-1; 1, b(a-1)\} \qquad (a, b > 1).$$

(Note that any imprimitive permutation group preserving a blocks of imprimitivity each of size b acts vertex-transitively on $K_{a \times b}$.)

If Γ is any diameter 2 distance-regular graph with intersection array

$$\{b_0, b_1; c_1, c_2\},$$

such that Γ is not a complete multipartite graph (so that $b_0 > c_2$), then the complement of Γ is also distance-regular, and has intersection array

$$\left\{ \frac{b_0 b_1}{c_2}, b_0 - c_2; 1, \frac{b_0 b_1}{c_2} - b_1 \right\}.$$

3

The Methods

We now describe the methods used to classify the transitive representations of rank at most 5 of the sporadic almost simple groups, and to analyse their associated generalized orbital digraphs. The methods for the analysis of these digraphs via their collapsed adjacency matrices are general, and are not restricted to sporadic groups or to low rank representations.

3.1 Permutation characters

Let G be a finite group. For a permutation representation of G on a finite set Ω, the *permutation character* π is the map $\pi : G \to \mathbb{C}$ such that, for $g \in G$, $\pi(g)$ is the number of points of Ω fixed by g. Now π is the character (trace map) of the natural matrix representation for G on the complex vector space $\mathbb{C}^\Omega$ (see for example [Isa76, p. 68]), and so π is a sum of complex irreducible characters for G. In particular the multiplicity of the trivial character 1_G in π is equal to the number of orbits of G in Ω (see [Isa76, Corollary 5.15]) and in particular π is a *transitive permutation character* (that is, the permutation character of a transitive permutation representation) precisely when 1_G has multiplicity 1 in π. In order to understand our arguments concerning permutation characters, the reader who is not familiar with the basic theory of (permutation) characters for finite groups is urged to consult the excellent reference [Isa76], especially Chapter 5. Note that permutationally isomorphic permutation representations of G have the same permutation character, but permutation representions which are not permutationally isomorphic may or may not have the same permutation character.

Improved versions of the ATLAS character tables (and power maps) for
all the sporadic simple groups and their automorphism groups are available in the computational group theory system GAP [Sch95]. We used
this system to obtain a list of possible permutation characters for transitive representations of rank at most 5 of these groups. More specifically,
there are several special properties that such a permutation character
must possess; for example it must be (rational) integer-valued and contain the trivial character 1_G exactly once. The list of properties of
interest to us is given in the statement below. (Here g^G denotes the
conjugacy class of g in G.)

Theorem 3.1 *Let π be the permutation character of a transitive representation of rank $r \leq 5$ of a finite group G, and let $g \in G$. Then*

  (a) $\pi = 1_G + \chi_2 + \cdots + \chi_r$, *where $\chi_2, \ldots, \chi_r$ are distinct nontrivial
      irreducible characters; in particular, π is multiplicity-free;*
  (b) $\pi(1)$ *divides* $|G|$;
  (c) $\pi(g)$ *is a non-negative integer;*
  (d) $\pi(1)$ *divides* $|g^G|.\pi(g)$;
  (e) $\pi(g^k) \geq \pi(g)$ *for all integers $k \geq 0$;*
  (f) *if $\pi(g) > 0$ then $\pi(1).|\langle g \rangle|$ divides $|G|$.*

Proof Properties (b) to (f) come from [Isa76, Theorem 5.18] and hold
for any transitive permutation character π of G.
Applying [Isa76, Corollaries 5.15 and 5.16], we see that either (a) holds,
or $r = 5$ and $\pi = 1_G + 2\chi$ for some nontrivial irreducible character χ.
Suppose the latter holds. The character χ has algebraic integer values
and, since π has rational integer values, χ too must have rational integer
values. Thus, each value of π is an odd rational integer, and so positive.
But then, as $\pi \neq 1_G$, the inner product (see [Isa76, pp. 20–21]) $[1_G, \pi]$ is
greater than 1, which contradicts the fact that 1_G has multiplicity 1 in
π.   □

3.2 Pseudo-permutation characters

We call a character π of a finite group G which satisfies (for all $g \in G$)
the properties (a) to (f) of Theorem 3.1 a *pseudo-permutation character*
for G (of rank at most 5). We first used GAP to compute a list of all
pseudo-permutation characters of rank at most 5 for the sporadic almost
simple groups. We then looked at each faithful such pseudo-permutation

character to determine if it really was a permutation character for the group G under consideration and, if so, worked to identify all permutation representations with that character. Usually the permutation characters corresponding to primitive representations of low rank of G were identified immediately using the information about maximal subgroups of G contained in the ATLAS. Some of the pseudo-permutation characters which were not permutation characters could be eliminated by proving that there was no suitable subgroup of G to be a point stabilizer, or by proving that the character value at some element of G was not a possible value for a permutation character evaluated at that element. Delicate analysis was often required in the cases where the character was shown (if it really was a permutation character) to be the permutation character of G acting (by right multiplication) on the right cosets of a subgroup H, where $H < M < G$, M a certain maximal subgroup of G. Such a representation would be imprimitive. In these cases it was often necessary to identify all subgroups H of M of the appropriate index and to examine the representation of G on the cosets of H for all of these subgroups H.

3.3 Constructing and analysing representations

Some of the permutation representations under consideration were originally analysed by hand, and this analysis is described with the corresponding representation.

Many other representations were constructed and analysed using the graph theory system GRAPE [Soi93b], which is a share library package in the group theory system GAP [Sch95]. GRAPE includes B.D. McKay's graph automorphism package *nauty* [McK90], L.H. Soicher's coset enumeration package *Enum*, and his program *Coladj* to compute collapsed adjacency matrices. The standalone version of the EnumColadj function in GRAPE was most useful. Its purpose is to produce collapsed adjacency matrices for the orbital digraphs of the transitive representation produced by a (successful) coset enumeration of the cosets of H in G, given a presentation for G and words generating H with respect to this presentation. The presentations we used come from [CNS88, CP92, HS95, Soi85, Soi87a, Soi88, Soi90, Soi91]. Many of these presentations are reproduced in the ATLAS (and in this book). The largest representation constructed and analysed this way was the degree 1545600 representation of Co_1 on the cosets of $3 \cdot Suz \colon 2$, and for this calculation we used a University of London Convex supercomputer. The reader who is unfamiliar with coset enumeration (also called the Todd-Coxeter method) should consult [Neu82] for a clear introduction.

Given a presentation for a group G, finding words generating a subgroup in a specified conjugacy class is as much an art as it is a science, but once found, it is usually easy to show that these words generate a subgroup in the correct conjugacy class. For example, a coset enumeration, when successful, will show that the subgroup H generated by these words has the desired index in G. Also, the calculation of the collapsed adjacency matrices for the orbital digraphs determines (among other things) the rank and subdegrees for the representation of G on the cosets of H. We shall give the presentations we have used explicitly, along with sets of words generating the appropriate subgroups. The systems Cayley, GAP and GRAPE were all useful in the determination of these words.

A small number of representations were constructed and analysed without using a presentation, but using the group theory system Cayley [Can84] (which has been superseded by the computer algebra system MAGMA [CP95]). In particular, we originally used Cayley to construct and analyse some of the imprimitive representations for M_{11} and M_{12}, for the initial study of the rank 5 representations of McL, and for the construction and analysis of the representations of Ru of rank at most 5. (We remark that Cayley, MAGMA and GAP contain libraries of matrix and permutation representations, including many for sporadic simple groups. See also [Wil].) Given some representation of a group G, Cayley was used to construct an appropriate point stabilizer H for the representation of G we wished to construct. For example, H could sometimes be found as the stabilizer of a subset of the points of the given representation for G, or as the normalizer of an appropriate subgroup of G. For imprimitive representations, H was usually a term in some composition series for the point stabilizer in some particular representation for G. After finding H, we then used the Cayley function `cosact image` to construct explicitly the desired permutation representation of G on the cosets of H. Next, we applied a simple Cayley version of our collapsed adjacency matrix program to calculate the collapsed adjacency matrices for the orbital digraphs for this representation.

The degree 9606125 permutation representation of Ly on the cosets of $3 \cdot McL : 2$ was constructed by Cooperman, Finkelstein, York and Tselman [CFYT94], and was kindly provided to us via the Institute for Experimental Mathematics, University of Essen.

We are also grateful to D.V. Pasechnik, who wrote a program in *Maple* [CGGLMW92] which calculated the collapsed adjacency matrices for all the orbital digraphs for the representations of Ly on the cosets of $G_2(5)$ and the Baby Monster group B on the cosets of $2^{.2}E_6(2)\colon 2$, given one collapsed adjacency matrix having distinct eigenvalues, for an orbital digraph, for each of these representations.

3.4 Computing with collapsed adjacency matrices

Throughout this section G is a transitive permutation group on a finite set Ω.

In this section we describe how to calculate the collapsed adjacency matrix for any generalized orbital digraph Γ for G. We also show how to use this collapsed adjacency matrix to determine whether Γ is connected, and if so its diameter; whether Γ is simple, and if so its girth; whether Γ is distance-regular, and whether G acts distance-transitively on Γ.

The methods described here are implemented in GRAPE. More advanced methods for computing collapsed adjacency matrices are discussed in [ILLSS95] and [LLS95].

3.4.1 Computing collapsed adjacency matrices

Let (Ω, E) be a generalized orbital digraph for G. Therefore E is the union of distinct G-orbitals, $E_{i_1}, \ldots, E_{i_s}$ say, and the collapsed adjacency matrix for (Ω, E) is just the sum of those for the orbital digraphs $(\Omega, E_{i_1}), \ldots, (\Omega, E_{i_s})$. Thus, we can determine the collapsed adjacency matrices for all generalized orbital digraphs for G from those for the orbital digraphs for G (given a fixed ordering of the suborbits for G).

We now show how to compute the collapsed adjacency matrices for the orbital digraphs for G. We suppose we are given permutation generators for G acting on Ω. We first calculate generators for the stabilizer G_α of a point $\alpha \in \Omega$. If we have calculated the permutation generators for G on Ω by a coset enumeration, then we already have words for generators of G_α (see [Neu82]). If the permutations generating G come from another source then we use standard methods from computational group theory [But91, Leo80] to calculate generators for G_α.

Next, we determine the orbits, $\Omega_1 = \{\alpha\}, \Omega_2, \ldots, \Omega_r$ say, of G_α on Ω, with respective representatives $\alpha_1 = \alpha, \alpha_2, \ldots, \alpha_r$ (see [But91, Chapter 7] for orbit algorithms). In the process, we determine a function

$$\text{suborb} : \Omega \to \{1, \ldots, r\}$$

(implemented as an array) such that, for $\beta \in \Omega$, $\text{suborb}(\beta) = j$ means that $\beta \in \Omega_j$. We also do a modified orbit calculation for the orbit (already known to be Ω) of G on Ω, in order to calculate a Schreier vector (see [But91, Chapter 7]) for this orbit. A Schreier vector allows us to calculate efficiently a word in the generators of G taking the point α to any given $\beta \in \Omega$. In particular, we calculate words $w_1, \ldots, w_r$, such that

$$\alpha^{w_1} = \alpha_1, \ldots, \alpha^{w_r} = \alpha_r.$$

Let $1 \leq i, k \leq r$. We now describe how to compute the kth row of the collapsed adjacency matrix A_i for the orbital digraph $\Gamma = (\Omega, E_i)$ corresponding to the suborbit Ω_i. Now $\Gamma(\alpha_k)$, the set of neighbours of α_k in Γ, is the image of $\Omega_i = \Gamma(\alpha)$ under (the permutation defined by) w_k. Thus, for $j = 1, \ldots, r$, we must calculate

$$A_i[k, j] := |\Gamma(\alpha_k) \cap \Omega_j| = |\Omega_i^{w_k} \cap \Omega_j|.$$

We can efficiently compute these $A_i[k, j]$ (for $j = 1, \ldots, r$) by executing the following steps:

  (i) for $j := 1, \ldots, r$: set $A_i[k, j] := 0$;
  (ii) for each $\beta \in \Omega_i$: set $j := \text{suborb}(\beta^{w_k})$, and then set $A_i[k, j] := A_i[k, j] + 1$.

After these steps have been executed, the kth row of A_i is correctly determined.

3.4.2 Applying collapsed adjacency matrices

As usual, G is a transitive permutation group on the finite set Ω, with $\alpha \in \Omega$. Let $A = (A[k, j])$ be the $r \times r$ collapsed adjacency matrix for a generalized orbital digraph $\Gamma = (\Omega, E)$ for G, with respect to a fixed ordering $\Omega_1 = \{\alpha\}, \Omega_2, \ldots, \Omega_r$ of the orbits of G_α.

Now define the *collapsed digraph*

$$\Sigma = \Sigma(A)$$

to have vertex set $\{1, \ldots, r\}$, with (k, j) an edge of Σ if and only if $A[k, j] > 0$.

Let $1 \leq k, j \leq r$. From the definition of A in Section 2.3, and the discussion following, it follows that (k, j) is an edge of Σ if and only if each element of Ω_k has a neighbour (in Γ) in Ω_j, and this holds if and only if at least one element in Ω_k has a neighbour in Ω_j. Since G_α is transitive on Ω_j, this also means that (k, j) is an edge of Σ if and only if each element of Ω_j is a neighbour in Γ of some element of Ω_k.

Some important properties of Γ can be read off the usually much smaller digraph Σ. Indeed, for the graphs in this book, Σ has at most five vertices.

Theorem 3.2 *Retaining the notation of this subsection, we have:*
(a) Γ *is simple if and only if* $(v, 1)$ *is an edge of* Σ *whenever* $(1, v)$ *is, but* $(1, 1)$ *is not an edge of* Σ.
(b) *Let* $1 \leq v \leq r$, $\beta \in \Omega_v$, *and let* n *be a non-negative integer. Then* α *is connected to* β *by a path of length* n *in* Γ *if and only if* 1 *is connected to* v *by a path of length* n *in* Σ. *(Thus, if* α *is connected to* β *by a path in* Γ, *then* $d_\Gamma(\alpha, \beta) = d_\Sigma(1, v)$.) *Furthermore,* Γ *is connected if and only if* Σ *is connected, in which case* $\mathrm{diam}(\Gamma) = \mathrm{diam}(\Sigma)$.

Proof (a) Since $\Omega_1 = \{\alpha\}$, the 'only if' part of (a) is obvious. Suppose that $(v, 1)$ is an edge of Σ whenever $(1, v)$ is, and $(1, 1)$ is not an edge of Σ. Suppose (α, γ) is an edge of Γ, and let $j := \mathrm{suborb}(\gamma)$ (that is, $\gamma \in \Omega_j$). Then $A[1, j] > 0$ and $(1, j)$ is an edge of Σ. Since $(1, 1)$ is not an edge of Σ, we have $j \neq 1$, that is, $\gamma \neq \alpha$, and so by the vertex-transitivity of Γ we can conclude that Γ contains no loops. Moreover, since $(1, j)$ is an edge, so also $(j, 1)$ is an edge of Σ, so $A[j, 1] > 0$. This implies that each element of Ω_j has a neighbour (in Γ) in $\Omega_1 = \{\alpha\}$. In particular, (γ, α) is an edge of Γ. Again, by the vertex-transitivity of Γ, we conclude that Γ is undirected. Thus Γ is simple.
(b) Suppose that $\beta_1, \beta_2, \ldots, \beta_{n+1}$ is a path of length n in Γ. Then this induces a path $v_1, v_2, \ldots, v_{n+1}$ in Σ, where for $i = 1, 2, \ldots, n + 1$, $\beta_i \in \Omega_{v_i}$. It follows that $d_\Sigma(v_1, v_{n+1}) \leq d_\Gamma(\beta_1, \beta_{n+1})$. Also, if Γ is connected, then so is Σ, and $\mathrm{diam}(\Gamma) \geq \mathrm{diam}(\Sigma)$. In the particular case where $\beta_1 = \alpha$, $\beta_{n+1} = \beta$, we have $v_1 = 1$, $v_{n+1} = v$, and $d_\Sigma(1, v) \leq d_\Gamma(\alpha, \beta)$.

Suppose now that 1 and v are connected by a path of length n in Σ. We shall show that α and β are connected by a path of length n in Γ, for every $\beta \in \Omega_v$. We do this by induction on n. If $n = 0$ then $v = 1$, $\beta = \alpha$, and the result holds. Suppose $n > 0$, that $1 = v_1, v_2, \ldots, v_{n+1} = v$ is a path of length n connecting 1 and v in Σ, and that $\beta \in \Omega_v$. By induction, there is a path of length $n - 1$ connecting α to γ for every $\gamma \in \Omega_{v_n}$. Since (v_n, v_{n+1}) is an edge of Σ, this implies that each element of $\Omega_{v_{n+1}} = \Omega_v$ is a neighbour in Γ of some element of Ω_{v_n}. Thus there is a path in Γ of length n from α to each $\beta \in \Omega_v$.

We thus have that $d_\Gamma(\alpha, \beta) \leq d_\Sigma(1, v)$, for $\beta \in \Omega_v$. Also, since Γ is vertex-transitive, we have that, if Σ is connected, then so is Γ, and $\mathrm{diam}(\Sigma) \geq \mathrm{diam}(\Gamma)$. This completes the proof of (b).   □

We note that, if $\Sigma = \Sigma(A)$ is connected, then its diameter $\mathrm{diam}(\Sigma) = \mathrm{diam}(\Gamma)$ is equal to the maximum $d_\Sigma(1, v)$, where v ranges over the vertices of Σ.

Now suppose that our generalized orbital digraph $\Gamma = (\Omega, E)$ is simple and connected of diameter d, and let $A = (A[k, j])$ and $\Sigma = \Sigma(A)$ be as before. Now Σ has diameter d, and as usual we let $\Sigma_i(1)$ denote the set of vertices of Σ at distance i from 1, for $i = 0, \ldots, d$. We see from the theorem above that

$$\bigcup_{v \in \Sigma_i(1)} \Omega_v = \Gamma_i(\alpha),$$

the set of vertices of Γ at distance i from α.

For $0 \leq i < d$ and $v \in \Sigma_i(1)$, define

$$b_i[v] := \sum_{w \in \Sigma_{i+1}(1)} A[v, w].$$

For $0 < i \leq d$ and $v \in \Sigma_i(1)$, define

$$c_i[v] := \sum_{w \in \Sigma_{i-1}(1)} A[v, w].$$

For $0 \leq i \leq d$ and $v \in \Sigma_i(1)$, define

$$a_i[v] := \sum_{w \in \Sigma_i(1)} A[v, w].$$

Suppose $\beta \in \Omega_v$, $v \in \Sigma_i(1)$, and so $d_\Gamma(\alpha, \beta) = i$. Then

$$b_i[v] = |\Gamma(\beta) \cap \Gamma_{i+1}(\alpha)|$$

(if $i < d$),

$$c_i[v] = |\Gamma(\beta) \cap \Gamma_{i-1}(\alpha)|$$

(if $i > 0$), and

$$a_i[v] = |\Gamma(\beta) \cap \Gamma_i(\alpha)|.$$

Theorem 3.3 *We retain the notation above. In particular, Γ is simple and connected and $d = \mathrm{diam}(\Gamma)$.*
(a) The graph Γ has no cycles if and only if, for all $i = 1, \ldots, d$ and all $v \in \Sigma_i(1)$, $c_i[v] \le 1$ and $a_i[v] = 0$.
Suppose that Γ contains a cycle and that m is the least positive integer i such that, for some $v \in \Sigma_i(1)$, $c_i[v] \ge 2$ or $a_i[v] \ge 1$. Define $g := 2m$ if $c_m[v] \ge 2$ for some $v \in \Sigma_m(1)$, and otherwise $g := 2m + 1$. Then Γ has girth g.
(b) Γ is distance-regular, with intersection array

$$\{b_0, b_1, \ldots, b_{d-1}; c_1, c_2, \ldots, c_d\},$$

if and only if for all $i = 0, \ldots, d$, and for all $v \in \Sigma_i(1)$, we have $b_i = b_i[v]$ (if $i < d$) and $c_i = c_i[v]$ (if $i > 0$).
(c) G acts distance-transitively on Γ if and only if $d = r - 1$, where r is the rank of G.

Proof (a) If Γ has no cycles then clearly the parameter restrictions $c_i[v] \le 1$ and $a_i[v] = 0$ hold. Suppose Γ has a shortest cycle C of length n. Since Γ is vertex-transitive, we may assume, without loss of generality, that α is a vertex in C. If $n = 2i$ is even then we must have $c_i[v] \ge 2$ for some $v \in \Sigma_i(1)$. If $n = 2i + 1$ is odd, then we must have $a_i[v] \ge 1$ for some $v \in \Sigma_i(1)$. This finishes the proof of the first statement of (a), and shows that g is a lower bound for the girth of Γ. Now suppose that $c_i[v] \ge 2$ for some $i > 0$ and $v \in \Sigma_i(1)$. Then there are two different paths in Γ from α to the same vertex in $\Gamma_i(\alpha)$, and so Γ must contain a cycle of length at most $2i$. If $a_i[v] > 0$ for some $i > 0$ and $v \in \Sigma_i(1)$, then this means that there are two (different) vertices in $\Gamma_i(\alpha)$ joined by an edge, and therefore Γ contains a cycle of length at most $2i + 1$. This shows that g is an upper bound for the girth of Γ, completing the proof of (a).
(b) This follows easily from the discussion preceding the statement of this theorem, together with the vertex-transitivity of Γ.
(c) This follows from the fact that G acts on Γ distance-transitively

if and only if the G-orbitals are precisely the (G-invariant) sets Γ_i of ordered pairs of vertices of Γ at distance i, for $i = 0, \ldots, d$.   □

It is worth pointing out that a simple connected vertex-transitive graph has no cycles if and only if the graph has at most two vertices.

In summary, let Γ be a generalized orbital digraph for a finite transitive permutation group G, such that the associated collapsed adjacency matrix for Γ is A, and the associated collapsed digraph is $\Sigma = \Sigma(A)$. Then Σ provides a simple tool for determining whether Γ is simple, whether Γ is connected, and if so for finding its diameter, and whether G acts on Γ distance-transitively. If Γ is simple and connected, then Σ together with A provide the tool-kit for determining the girth of Γ, and whether Γ is distance-regular.

3.5 A lemma for certain imprimitive groups

The result below will be used to eliminate from consideration certain imprimitive representations which have rank greater than 5.

Lemma 3.4 *Let G be a transitive imprimitive permutation group on a finite set Ω, and let $\Phi = \{B^g \mid g \in G\}$ be a system of blocks of imprimitivity preserved by G. Suppose that G acts 2-transitively on Φ, and that, for $B \in \Phi$, the actions of G_B (the setwise stabilizer of B in G) on B and $\Phi \setminus \{B\}$ are permutationally isomorphic. Then the action of G on Ω is permutationally isomorphic to its action on the set of ordered pairs of distinct elements of Φ; in particular, if $|\Phi| \geq 4$ then G has rank at least 7 in its action on Ω.*

Proof Let $\alpha \in B$ and $H = G_B$. Then by assumption, $G_\alpha = H_\alpha = H_{B'} = G_{B,B'}$, for some $B' \in \Phi \setminus \{B\}$. Thus the actions of G on Ω and on the set of ordered pairs of distinct elements of Φ are permutationally isomorphic. If $|\Phi| \geq 4$ then the following seven sets are invariant under $G_\alpha = H_\alpha = G_{B,B'}$: $\{(B, B')\}$, $\{(B', B)\}$, $\{(B, C) \mid C \neq B, B'\}$, $\{(B', C) \mid C \neq B, B'\}$, $\{(C, B) \mid C \neq B, B'\}$, $\{(C, B') \mid C \neq B, B'\}$, $\{(C, C') \mid C, C' \neq B, B'\}$, in which case G has rank at least 7 on Ω.   □

4

The Individual Groups

We now handle individually each sporadic almost simple group G having a faithful pseudo-permutation character of rank at most 5, and list each such character π found by our computer search. The numbering of the irreducible characters χ_i agrees with that in the GAP version of the ATLAS character tables, which agrees with the printed ATLAS [CCNPW85] in the case of simple groups. We examine each of these pseudo-permutation characters in turn to determine if it is the character of some permutation representation of G, and, if so, we describe all such representations (up to permutational isomorphism). Specifically, for each representation of rank up to 5, we provide

- the irreducible constituents of the permutation character, as well as the degrees of these constituents,
- a point stabilizer,
- (for rank > 2) the collapsed adjacency matrices for the nontrivial orbital digraphs, and
- (for rank > 3) the non-complete distance-regular generalized orbital graphs.

Note that the information on rank 2 and 3 graphs which is suppressed in this chapter is given in a general way in Section 2.5.

We also give presentations for many sporadic almost simple groups. Specifically, let K be a sporadic simple group, other than Ru, Ly, and B, such that K has a faithful permutation representation of rank at most 5. Then we shall give an explicit presentation for $G = K$ or $G = \mathrm{Aut}\,(K)$, or both. We shall also give collections of words generating various subgroups of these groups G, which include all those we

29

used to construct, via coset enumeration, the representations of rank at
most 5 giving rise to the collapsed adjacency matrices for K and Aut (K)
given in this book. (If $K \not\cong$ Aut (K), and there is a representation ρ of
Aut $(K) = K.2$ on the cosets of $L.2 \not\leq K$, such that the the restriction
of ρ to K on the cosets of $L \leq K$ has the same rank (≤ 5) as does ρ,
then we sometimes give a presentation and subgroup generators for just
one of these representations, as they both give rise to the same (orbital
digraphs and) collapsed adjacency matrices.) For readers familiar with
GRAPE, these presentations and subgroup generators will help them to
use GRAPE to construct and study the generalized orbital digraphs of
rank at most 5 for K and Aut (K).

The reader who wants more information on the sporadic groups than
is given in this book is urged to consult the ATLAS [CCNPW85], the
ATLAS of Brauer Characters [JLPW95], which contains many references
on the sporadic groups, [Gor82], which contains interesting historical
information about these groups, [CS88], and [Asc94].

4.1 Notation and conventions

In each section for the individual groups, G always denotes the sporadic
almost simple group under consideration. For π a permutation character
of G on a set Ω of $\pi(1)$ points, we always let H denote the point stabilizer
G_α, where $\alpha \in \Omega$. We give the degrees of the irreducible constituents of
π as the summands of $\pi(1)$.

For the purposes of constructing the collapsed adjacency matrices, we
order the suborbits in non-decreasing order of their lengths, with the first
suborbit $\Omega_1 = \{\alpha\}$ corresponding to the trivial orbital. In the case of
repeated subdegrees greater than 1 an ordering of the suborbits of equal
length is chosen arbitrarily. We print the collapsed adjacency matrices
for the nontrivial orbital digraphs in the order of their corresponding
suborbits. Note that the unique non-zero entry in the first row of a
collapsed adjacency matrix for an orbital digraph Γ is the corresponding
suborbit length as well as the (out) valency of Γ.

If the rank r of G on Ω is greater than 3, then after giving the collapsed adjacency matrices $A_2, A_3, \ldots, A_r$ for the nontrivial orbital digraphs (which correspond to suborbits $\Omega_2, \Omega_3, \ldots, \Omega_r$), we describe each
non-complete distance-regular generalized orbital graph $\Gamma = (\Omega, E)$ for

this representation. Such a graph $\Gamma = (\Omega, E)$ is denoted

$$\Gamma\{i_1, \ldots, i_s\},$$

where $i_1, \ldots, i_s$ are distinct, and this notation means that E is the (disjoint) union of the orbitals corresponding to the suborbits $\Omega_{i_1}, \ldots, \Omega_{i_s}$. Recall that this implies that the collapsed adjacency matrix for Γ is

$$A_{i_1} + \cdots + A_{i_s}$$

(Proposition 2.2). For each non-complete distance-regular generalized orbital graph Γ for G, we also give its intersection array

$$\iota = \iota(\Gamma),$$

unless Γ is complete multipartite or the complement of a diameter 2 distance-regular graph already described, and we indicate when G acts distance-transitively on Γ. Again we note that this information is given for representations of rank 2 and 3 in a general way in Section 2.5.

Each transitive group on n points has the complete graph K_n as a generalized orbital graph (obtained from the union of all nontrivial orbitals), and of course the graph K_n has a distance-transitive action by its automorphism group S_n.

Each transitive imprimitive group having a blocks of imprimitivity of size b has the complete multipartite graph $K_{a\times b}$ as a generalized orbital graph (obtained from the union of the orbitals corresponding to the suborbits not in the block containing the fixed point α), which has a distance-transitive action by its automorphism group $S_b \operatorname{wr} S_a$.

For $1 \leq k \leq m$, the *Johnson graph* $J(m, k)$ is the graph whose vertex-set is the set of all k-subsets of an m-set, with two vertices joined by an edge if and only if their intersection has size $k - 1$. The symmetric group S_m, acting naturally on the k-subsets of an m-set, acts distance-transitively on $J(m, k)$.

We shall also encounter certain well-known Taylor graphs. A *Taylor graph* is a distance-regular graph with intersection array of the form $\{k, \mu, 1; 1, \mu, k\}$. More information on Taylor graphs can be found in [BCN89].

Some of the groups we describe act on t-designs. Let $v \geq k \geq t \geq 1$ and $\lambda \geq 1$ be integers. Then a $t - (v, k, \lambda)$ *design* consists of a v-set X together with a collection $\mathcal{B}$ of k-subsets of X, such that every t-subset of X is contained in exactly λ elements of $\mathcal{B}$. The elements of X are called *points*, and the elements of $\mathcal{B}$ are called *blocks*. A *Steiner system* $S(t, k, v)$ is simply a $t - (v, k, 1)$ design.

Our notation for group structures and conjugacy classes follows the AT-LAS. We also make extensive use of the maximal subgroup information in the ATLAS, often without reference, as well as the ATLAS character tables and permutation character information (although our calculations provide a check on much of the permutation character information we use).

4.1.1 Notation for group presentations

In our compact notation for a group presentation, we give many of the generators and relations via a Coxeter graph. Usually all the generators, but not all the relations, are given this way.

Each node x of a Coxeter graph specifies a generator x and the relation $1 = x^2$. If two distinct nodes x, y in a Coxeter graph are joined by an edge with label m, then this denotes the relation $1 = (xy)^m$. (All the edges in our Coxeter graphs are labelled (with positive integers), although the usual convention is to omit the label 3.) If two distinct nodes x, y in a Coxeter graph are not joined by an edge, then this denotes the relation $1 = (xy)^2$. We specify a Coxeter graph Φ by a set of paths in Φ, each containing at least one edge, and which together contain all the edges of Φ. A typical path is of the form

$$xmynz\ldots,$$

which means that x is joined to y with an edge labelled m, y is joined to z with an edge labelled n, and so on.

For example, the graph specified by the two paths

$$a3b3c3d3a3c, b3d$$

is the complete graph on a, b, c, d, with all edge-labels 3. The Coxeter graph

is denoted by

$$a3b5c,$$

which also denotes the presentation

$$\langle a, b, c \mid 1 = a^2 = b^2 = c^2 = (ab)^3 = (bc)^5 = (ac)^2 \rangle$$

for $2 \times A_5$.

Relations that are not specified in a Coxeter graph are given in the usual way. For example,

$$A_5 \cong \langle a3b5c \mid 1 = (abc)^5 \rangle.$$

If w_1, w_2 are words, then $w_1^{w_2}$ means $w_2^{-1} w_1 w_2$. Generators that are not specified in a Coxeter graph are listed before the Coxeter graph paths. Finally, relations given in square brackets are redundant, but may be helpful for the purpose of coset enumeration. Thus, in the presentation

$$\langle a3b5c3d \mid 1 = (abc)^5 [= (bcd)^5] \rangle$$

(for $L_2(11)$), the relation $1 = (bcd)^5$ holds, but is a consequence of the other relations.

4.2 The Mathieu group M_{11}

A presentation for M_{11} is

$$\langle a3b5c3d3e4c \mid a = (ce)^2, [1 = (abc)^5 = (bcd)^5] \rangle$$

(see [Soi87b]), in which

$$M_{10} \cong \langle b, c, d, abcbae \rangle > \langle b, c, d, d^{abcbae} \rangle \cong A_6,$$

$$L_2(11) \cong \langle a, b, c, d \rangle,$$

$$M_9 : 2 \cong \langle a, b, d, e, (decbcdcba)^{cbc} \rangle > \langle ab, de, (decbcdcba)^{cbc} \rangle \cong 3^2 : 8,$$

and

$$S_5 \cong \langle a, b, c, e \rangle.$$

4.2.1 rank 2

(i) $\pi = \chi_1 + \chi_2$, of degree $11 = 1 + 10$.

This character corresponds to the unique 2-transitive representation of G of degree 11. A point stabilizer is $H \cong M_{10}$.

(*ii*) $\pi = \chi_1 + \chi_5$, of degree $12 = 1 + 11$.

This character corresponds to the unique 2-transitive representation of G of degree 12. A point stabilizer is $H \cong L_2(11)$.

4.2.2 rank 3

(*iii*) $\pi = \chi_1 + \chi_2 + \chi_5$, of degree $22 = 1 + 10 + 11$.

It follows from the maximal subgroup structure of G that $H < L < G$, where $L \cong M_{10}$ is a point stabilizer in the representation (*i*) above. Hence G preserves a system of imprimitivity $\Sigma := B^G$ where $B := \alpha^L$, and $H \cong A_6$ is transitive on $\Sigma \setminus \{B\}$. Let $B' \in \Sigma \setminus \{B\}$ and $\beta \in B'$. Then $H_{B'} \cong 3^2{:}4$. Let $g \in H_{B'}$ be an element of order 4. Then g fixes $\pi(g) = 1 + 2 - 1 = 2$ points of Ω, and hence g fixes only the two points of B. It follows that $H_{B'}$ is transitive on B', whence H is transitive on $\Omega \setminus B$. Thus G has a unique imprimitive rank 3 permutation representation of degree 22, and its character must therefore be π. The subdegrees are 1, 1, 20.

The collapsed adjacency matrices for the (nontrivial) orbital digraphs are

$$\begin{pmatrix} 0 & 1 & 0 \\ 1 & 0 & 0 \\ 0 & 0 & 1 \end{pmatrix}$$

$$\begin{pmatrix} 0 & 0 & 20 \\ 0 & 0 & 20 \\ 1 & 1 & 18 \end{pmatrix}.$$

(*iv*) $\pi = \chi_1 + \chi_2 + \chi_8$, of degree $55 = 1 + 10 + 44$.

There is a unique class of maximal subgroups of G such that a representative has index 55, and the associated permutation character is π. A point stabilizer is $H \cong M_9{:}2$, the stabilizer of an unordered pair of points in the representation in (*i*).

The subdegrees are 1, 18, 36, and the collapsed adjacency matrices are

$$\begin{pmatrix} 0 & 18 & 0 \\ 1 & 9 & 8 \\ 0 & 4 & 14 \end{pmatrix}$$

$$\begin{pmatrix} 0 & 0 & 36 \\ 0 & 8 & 28 \\ 1 & 14 & 21 \end{pmatrix}.$$

If there were also an imprimitive permutation representation with this character then, again by examining the maximal subgroups of G, a point stabilizer H would be a subgroup of index 5 in a subgroup M_{10}, but M_{10} has no subgroup of this index.

4.2.3 rank 4

(v) $\pi = \chi_1 + \chi_2 + \chi_5 + \chi_8$, of degree $66 = 1 + 10 + 11 + 44$.

There is a unique class of maximal subgroups of G of index 66, and the associated permutation character is π. A point stabilizer is $H \cong S_5$, the stabilizer of an unordered pair of points in the representation in (ii).

The subdegrees are 1, 15, 20, 30, and the collapsed adjacency matrices are

$$\begin{pmatrix} 0 & 15 & 0 & 0 \\ 1 & 0 & 4 & 10 \\ 0 & 3 & 3 & 9 \\ 0 & 5 & 6 & 4 \end{pmatrix}$$

$$\begin{pmatrix} 0 & 0 & 20 & 0 \\ 0 & 4 & 4 & 12 \\ 1 & 3 & 10 & 6 \\ 0 & 6 & 4 & 10 \end{pmatrix}$$

$$\begin{pmatrix} 0 & 0 & 0 & 30 \\ 0 & 10 & 12 & 8 \\ 0 & 9 & 6 & 15 \\ 1 & 4 & 10 & 15 \end{pmatrix}.$$

The non-complete distance-regular generalized orbital graphs are

$$\Gamma\{3\} \cong J(12,2), \quad \iota = \{20,9;1,4\} \quad \text{(and complement)}.$$

If there were also an imprimitive permutation representation with character π, then H would be a subgroup of index 6 in a subgroup M_{10}, but M_{10} has no subgroup of this index.

(vi) $\pi = \chi_1 + \chi_2 + \chi_8 + \chi_{10}$, of degree $110 = 1 + 10 + 44 + 55$.

It follows from the maximal subgroup structure of G that $H < L < G$, where either $L \cong M_{10}$ of index 11, or $L \cong M_9 : 2$ of index 55. In the former case, G preserves a partition Σ of Ω into eleven blocks of size 10, and the stabilizer $L = G_B$ of $B \in \Sigma$ induces permutationally isomorphic actions on B and $\Sigma \backslash \{B\}$, since M_{10} has a unique representation of degree 10. By Lemma 3.4, this action of G has rank at least 7. Therefore we have $H < L \cong M_9 : 2 \cong 3^2 : Q_8.2$. A Sylow 2-subgroup S of L is a Sylow 2-subgroup of G and is semidihedral of order 16,

$$S = \langle x, y \mid x^8 = y^2 = 1, x^y = x^3 \rangle$$

see [Gor68, p.487 and p.191], From the character table of G, we find that $\pi(g) = 2$ for each element $g \in G$ of order 8. Therefore H contains an element x of order 8, and it follows that $H = O_3(L).\langle x \rangle \cong 3^2 : 8$. The action of G on Σ is permutationally isomorphic to its action on the set of unordered pairs of elements from the set $\Delta := \{1, 2, \ldots, 11\}$ on which G acts naturally. We may identify Σ with this set so that L is the stabilizer of the pair $\{1, 2\}$, and then the orbits of L on Σ are $\{\{1, 2\}\}$, $\Sigma_1 := \{\{i, j\} \mid i \leq 2 < j\}$, and $\Sigma_2 := \{\{i, j\} \mid i, j > 2\}$, of lengths 1, 18 and 36 respectively. Clearly $O_3(L)$ has six orbits of length 9 on Σ, and hence has twelve orbits of length 9 on Ω. The permutation character π' for the action of G on Σ is the character in (iv). Since $\pi'(x) = 1$, x fixes exactly one element of Σ and hence H is transitive on Σ_1. Since $\pi'(x^2) = 3$ and $\pi(x^2) = 2$, x^2 fixes three elements of Σ, but fixes only two points of Ω. It follows that H is transitive on both Σ_2 and the set of points of Ω contained in elements of Σ_1. Finally, since $\pi(x^4) = 6$, x^4 fixes precisely the six points of Ω contained in the three elements of Σ fixed by x^2. In particular, x^4 fixes no points contained in elements of Σ_2. Hence H is transitive on the set of points of Ω contained in elements of Σ_2. Thus the imprimitive action of G on Ω has rank 4, with subdegrees 1, 1, 36, 72.

Thus we have described, up to permutational isomorphism, a unique imprimitive rank 4 representation of M_{11}. The collapsed adjacency matrices are

$$\begin{pmatrix} 0 & 1 & 0 & 0 \\ 1 & 0 & 0 & 0 \\ 0 & 0 & 1 & 0 \\ 0 & 0 & 0 & 1 \end{pmatrix}$$

$$\begin{pmatrix} 0 & 0 & 36 & 0 \\ 0 & 0 & 36 & 0 \\ 1 & 1 & 18 & 16 \\ 0 & 0 & 8 & 28 \end{pmatrix}$$

$$\begin{pmatrix} 0 & 0 & 0 & 72 \\ 0 & 0 & 0 & 72 \\ 0 & 0 & 16 & 56 \\ 1 & 1 & 28 & 42 \end{pmatrix}.$$

The only non-complete distance-regular generalized orbital graph is

$$\Gamma\{3,4\} \cong K_{55\times2}.$$

4.2.4 rank 5

G has no pseudo-permutation character of rank 5.

4.3 The Mathieu group M_{12}

A presentation for M_{12} is

$$\langle b4f3a3b5c3d3e4c, e6f \mid a = (ce)^2, d = (bf)^2, 1 = (adef)^3 = (bcef)^6 \rangle$$

(discovered jointly by A.J.E. Ryba and L.H. Soicher; see [Soi88]), in which

$$M_{11} \cong \langle a, b, c, d, e \rangle > \langle a, b, c, d \rangle \cong L_2(11),$$

and

$$M_{10}: 2 \cong \langle b, c, d, f, abcbae \rangle$$

containing

$$\langle b, c, d, fabcbae \rangle \cong PGL_2(9) \quad \text{and} \quad \langle b, c, d, f, d^{abcbae} \rangle \cong S_6.$$

We also have

$$L_2(11) \cong \langle abc, bcd, aefd \rangle \quad \text{(maximal)},$$

and

$$M_9: S_3 \cong \langle a, b, d, e, f \rangle.$$

A presentation for $M_{12}:2$ is obtained from the presentation for M_{12} by adjoining a generator t, and the relations

$$1 = t^2 = a^t d = b^t c = e^t f.$$

In this $M_{12}:2$, we have

$$\langle a, b, c, d, t \rangle \cong L_2(11):2 \cong \langle abc, bcd, aefd, t \rangle,$$

but note that these two subgroups are in different conjugacy classes of $M_{12}:2$.

4.3.1 rank 2

(i) $\pi = \chi_1 + \chi_i$, where $i = 2$ or 3, of degree $12 = 1 + 11$.

These characters correspond to the two equivalent, but not permutationally isomorphic, 2-transitive representations of G of degree 12, each having point stabilizer isomorphic to M_{11}. An equivalence between these representations, which interchanges them, is induced by an outer automorphism of G.

4.3.2 rank 3

(ii) $\pi = \chi_1 + \chi_i + \chi_7$, where $i = 2$ or 3, of degree $66 = 1 + 11 + 54$.

By examining the indices of maximal subgroups of G, it is clear that there is no imprimitive permutation representation of G of degree 66. It follows that the above representations correspond to the two classes of maximal subgroups of G having a representative of index 66, and these representations are interchanged by an outer automorphism of G. A point stabilizer is $M_{10}:2$, the stabilizer of an unordered pair of points in the representation with character $\chi_1 + \chi_i$ in (i).

Thus we have described two equivalent, but not permutationally isomorphic, primitive rank 3 representations of M_{12}. The subdegrees are 1, 20, 45, and the collapsed adjacency matrices are

$$\begin{pmatrix} 0 & 20 & 0 \\ 1 & 10 & 9 \\ 0 & 4 & 16 \end{pmatrix}$$

$$\begin{pmatrix} 0 & 0 & 45 \\ 0 & 9 & 36 \\ 1 & 16 & 28 \end{pmatrix}.$$

4.3.3 rank 4

(iii) $\pi = \chi_1 + \chi_i + \chi_7 + \chi_{11}$, where $i = 2$ or 3, of degree $132 = 1 + 11 + 54 + 66$.

From the table of maximal subgroups of G, it follows that $H < L < G$, where $m := |G : L|$ is either 12 or 66, and G preserves a system of imprimitivity $\Sigma := B^G$ where $|\Sigma| = m$, $B := \alpha^L$, and $L = G_B$. The permutation character π' of G in its action on Σ has degree m and is contained in π. Suppose first that $m = 12$. Then $\pi' = \chi_1 + \chi_i$, a character in (i), and $L \cong M_{11}$, which has a unique permutation representation of degree 11. It follows that H is the stabilizer of an ordered pair of distinct points in a permutation representation in (i), which implies by Lemma 3.4 that G has rank at least 7 on Ω.

This contradiction shows that $m = 66$. So $\pi' = \chi_1 + \chi_i + \chi_7$, a character in (ii), $L \cong M_{10} : 2 \cong \mathrm{Aut}\,(S_6)$, and H is one of the three subgroups of L of index 2. By Lemma 3.4 it follows that $H \not\cong M_{10}$. Now there is an element $x \in G$ of order 8, with $\pi(x) = 2$, and hence H contains a conjugate of x. We may assume that $x \in H$. Since S_6 contains no element of order 8, it follows that $H \cong PGL_2(9)$.

It follows from (ii) above that the action of G on Σ is permutationally isomorphic to the rank 3 action of G on the set of 66 unordered pairs from a set $\Delta := \{1, 2, \ldots, 12\}$ (on which G acts with associated permutation character $\chi_1 + \chi_i$). We may make this identification so that the orbits of L on Σ are $\{B\} = \{1, 2\}$, $\Sigma_1 := \{\,\{k, l\} \mid 1 \le k \le 2, l > 2\,\}$, and $\Sigma_2 := \{\,\{k, l\} \mid 2 < k < l\,\}$, of lengths 1, 20, and 45 respectively. Let $B_j \in \Sigma_j$ and let β_j be a point of Ω in B_j, for $j = 1, 2$. Now $H_{B_2} \cong D_{16}$ is a Sylow 2-subgroup of H, and so we may assume that the element x fixes B_2 setwise. Since x fixes only two points of Ω, and x fixes α, it follows that x interchanges the two points of B_2. Thus H is transitive on the 90 points of Ω lying in blocks in Σ_2.

We claim that all elements of H of order 4 fix no points of Ω in blocks of Σ_1. Since $H_{B_2} \cong D_{16}$ is a Sylow 2-subgroup of H, and since the two elements of order 4 in H_{B_2}, namely x^2 and x^6, are conjugate in H_{B_2}, it is sufficient to consider x^2. It follows from the character table for G that there are two conjugacy classes of elements of order 4, fixing zero or four points of Ω respectively. Thus x^2, which lies in H_{β_2}, fixes precisely the four points of Ω in $B \cup B_2$. In particular, x^2 fixes no points of Ω in any block in Σ_1.

Now consider the action of H on Σ_1. The subgroup L_{B_1} is the stabilizer in G of three points of Δ which, without loss of generality, we may take to be the points 1, 2, 3. Now H is transitive on $\Delta \setminus \{1,2\}$, and $H_3 \cong 3^2{:}8$ (see ATLAS, p.4), so we have $H_{B_1} \leq H_3 \cong 3^2{:}8$. Since $\pi'(x) = 1 + 1 + 0 = 2$, x fixes no block of Σ_1 setwise, and hence H_{B_1} contains no element of order 8, so $H_{B_1} \neq H_3$. It follows that H is transitive on Σ_1 and $H_{B_1} \cong 3^2{:}4$. From the previous paragraph we know that elements of order 4 in H_{B_1} do not fix any points in B_1. Hence $H_{\beta_1} \cong 3^2{:}2$ and H is transitive on the 40 points of Ω in blocks of Σ_1. Thus the imprimitive action of G on Ω has rank 4, with subdegrees 1, 1, 40, 90.

Note that we have described two equivalent, but not permutationally isomorphic, representations of M_{12}. An outer automorphism of M_{12} induces an equivalence which interchanges the two permutation representations arising in case (ii), and hence interchanges the two permutation representations in this case also. The collapsed adjacency matrices for each representation are

$$\begin{pmatrix} 0 & 1 & 0 & 0 \\ 1 & 0 & 0 & 0 \\ 0 & 0 & 1 & 0 \\ 0 & 0 & 0 & 1 \end{pmatrix}$$

$$\begin{pmatrix} 0 & 0 & 40 & 0 \\ 0 & 0 & 40 & 0 \\ 1 & 1 & 20 & 18 \\ 0 & 0 & 8 & 32 \end{pmatrix}$$

$$\begin{pmatrix} 0 & 0 & 0 & 90 \\ 0 & 0 & 0 & 90 \\ 0 & 0 & 18 & 72 \\ 1 & 1 & 32 & 56 \end{pmatrix}.$$

The only non-complete distance-regular generalized orbital graph is

$$\Gamma\{3,4\} \cong K_{66\times 2}.$$

4.3.4 rank 5

(iv) $\pi = \chi_1 + \chi_2 + \chi_3 + \chi_7 + \chi_i$, where $i = 8, 9$, or 10, of degree $132 = 1 + 11 + 11 + 54 + 55$. (We shall show that it is only when $i = 8$ that we get a permutation character.)

From the table of maximal subgroups of G, it follows that $H < L < G$, where $m := |G : L|$ is either 12 or 66, and G preserves a system of imprimitivity $\Sigma := B^G$ where $|\Sigma| = m$, $B := \alpha^L$, and $L = G_B$. The assumption $m = 12$ leads to a contradiction by the same argument as in case (iii) above. So $m = 66$. Let π' be the permutation character of G in its action on Σ. So $\pi' = \chi_1 + \chi_j + \chi_7$ with $j = 2$ or 3, a character in (ii), $L \cong M_{10} : 2 \cong \mathrm{Aut}\,(S_6)$, and H is one of the three subgroups of L of index 2. By Lemma 3.4 it follows that $H \not\cong M_{10}$. Also, by (iii) above, we know that H is not $PGL_2(9)$ since in that case we showed that the permutation representation has rank 4. Therefore $H \cong S_6$.

Let $x \in G$ be an element of order 8. Then $\pi(x) = 2$, if either $i = 9$ and x is in class $8B$, or $i = 10$ and x is in class $8A$. Since H contains no element of order 8, it follows that $i = 8$. Then it follows that $\pi(y) = 4$ for every element y of order 4, and hence H contains elements of order 4 from both class $4A$ and class $4B$ of G, and each of them fixes four points of Ω, two in B and two in one other block of Σ.

It follows from (ii) above that the action of G on Σ is permutationally isomorphic to the rank 3 action of G on the set of 66 unordered pairs from a set $\Delta := \{1, 2, \ldots, 12\}$ (on which G acts with associated permutation character $\chi_1 + \chi_j$). As in (iii) we may make this identification so that the orbits of L on Σ are $\{B\} = \{1, 2\}$, $\Sigma_1 := \{\,\{k, l\} \mid 1 \leq k \leq 2, l > 2\,\}$, and $\Sigma_2 := \{\,\{k, l\} \mid 2 < k < l\,\}$, of lengths 1, 20, and 45 respectively. Let $B_k \in \Sigma_k$ and let β_k be a point of Ω in B_k, for $k = 1, 2$. Since $H \not\subseteq G_{\{1,2\}} \cong M_{10}$, it follows that H is transitive on both Σ_1 and Σ_2. Suppose that H is intransitive on the 40 points of Ω in the blocks of Σ_1. Then H has two orbits of length 20 on these points, say $\Omega_{1,1}$ and $\Omega_{1,2}$. An element y of order 4 lying in $A_6 \subseteq H$ fixes four points of Δ including 1 and 2, and hence fixes four blocks of Σ_1. It must therefore fix four points of each of $\Omega_{1,1}$ and $\Omega_{1,2}$. This contradicts the fact that y fixes only four points of Ω. Thus H is transitive on the 40 points of Ω lying in the blocks of Σ_1. If H were transitive on the 90 points of Ω lying in the blocks of Σ_2, then H would have rank 4 in its action on Ω, and we know that this is not the case from the previous subsection. Hence H has rank 5 on Ω, with subdegrees 1, 1, 40, 45, 45.

Note that we have described two equivalent, but not permutationally isomorphic, permutation representations of G with character $\pi = \chi_1 + \chi_2 + \chi_3 + \chi_7 + \chi_8$. An equivalence which interchanges these two representations is induced by an outer automorphism of M_{12}. The collapsed adjacency matrices for each representation are

$$
\begin{pmatrix}
0 & 1 & 0 & 0 & 0 \\
1 & 0 & 0 & 0 & 0 \\
0 & 0 & 1 & 0 & 0 \\
0 & 0 & 0 & 0 & 1 \\
0 & 0 & 0 & 1 & 0
\end{pmatrix}
$$

$$
\begin{pmatrix}
0 & 0 & 40 & 0 & 0 \\
0 & 0 & 40 & 0 & 0 \\
1 & 1 & 20 & 9 & 9 \\
0 & 0 & 8 & 16 & 16 \\
0 & 0 & 8 & 16 & 16
\end{pmatrix}
$$

$$
\begin{pmatrix}
0 & 0 & 0 & 45 & 0 \\
0 & 0 & 0 & 0 & 45 \\
0 & 0 & 9 & 18 & 18 \\
1 & 0 & 16 & 20 & 8 \\
0 & 1 & 16 & 8 & 20
\end{pmatrix}
$$

$$
\begin{pmatrix}
0 & 0 & 0 & 0 & 45 \\
0 & 0 & 0 & 45 & 0 \\
0 & 0 & 9 & 18 & 18 \\
0 & 1 & 16 & 8 & 20 \\
1 & 0 & 16 & 20 & 8
\end{pmatrix}.
$$

The only non-complete distance-regular generalized orbital graph is

$$\Gamma\{3,4,5\} \cong K_{66\times 2}.$$

(v) $\pi = \chi_1 + \chi_2 + \chi_3 + \chi_8 + \chi_{11}$, of degree $144 = 1 + 11 + 11 + 55 + 66$.

There is only one conjugacy class of maximal subgroups of G of index 144, and the associated permutation character is not equal to this character π (see the ATLAS, p.33). It follows that $H < L < G$, where $m := |G : L| = 12$, and G preserves a system of imprimitivity $\Sigma := B^G$ where $|\Sigma| = m$, $B := \alpha^L$, and $L = G_B$. Also the permutation character π' of G in its action on Σ is equal to $\chi_1 + \chi_i$, where $i = 2$ or

3, and $L \cong M_{11}$, $H \cong L_2(11)$. The group L has a unique conjugacy class of subgroups of index 12, while G has two classes of subgroups of index 12, and (examining the associated permutation characters) taking L to belong to the first of these classes, L acts transitively on the set of cosets of the second class. It follows that H is the intersection of L and a subgroup M from the second conjugacy class of subgroups of G of index 12. Thus G preserves a second system of imprimitivity on Ω, $\Theta := C^G$, where $|\Theta| = 12$, $M = G_C$, $C = \alpha^M$, and $H = L \cap M$. Let $B' \in \Sigma \setminus \{B\}$. Now H has two representations of degree 11, and so the orbits of $H_{B'} \cong A_5$ in B (and hence in B') have lengths either 1, 1, 10, or 1, 5, 6, and hence the orbits of H in Ω have lengths either 1, 11, 11, 11, 110 or 1, 11, 11, 55, 66. In particular G has rank 5 on Ω, and we now determine the subdegrees. The subgroup H is normalised by an outer automorphism of G (see ATLAS), and $H.2 \cong PGL_2(11)$ permutes the H-orbits in Ω. If H had three orbits of length 11, then $H.2$ would fix one of them setwise, but $H.2$ has no subgroup of index 11. Hence G is imprimitive of rank 5 with subdegrees 1, 11, 11, 55, 66.

Thus we have described, up to permutational isomorphism, a unique imprimitive rank 5 representation with this permutation character. The collapsed adjacency matrices are

$$
\begin{pmatrix}
0 & 11 & 0 & 0 & 0 \\
1 & 10 & 0 & 0 & 0 \\
0 & 0 & 0 & 5 & 6 \\
0 & 0 & 1 & 4 & 6 \\
0 & 0 & 1 & 5 & 5
\end{pmatrix}
$$

$$
\begin{pmatrix}
0 & 0 & 11 & 0 & 0 \\
0 & 0 & 0 & 5 & 6 \\
1 & 0 & 10 & 0 & 0 \\
0 & 1 & 0 & 4 & 6 \\
0 & 1 & 0 & 5 & 5
\end{pmatrix}
$$

$$
\begin{pmatrix}
0 & 0 & 0 & 55 & 0 \\
0 & 0 & 5 & 20 & 30 \\
0 & 5 & 0 & 20 & 30 \\
1 & 4 & 4 & 22 & 24 \\
0 & 5 & 5 & 20 & 25
\end{pmatrix}
$$

$$\begin{pmatrix} 0 & 0 & 0 & 0 & 66 \\ 0 & 0 & 6 & 30 & 30 \\ 0 & 6 & 0 & 30 & 30 \\ 0 & 6 & 6 & 24 & 30 \\ 1 & 5 & 5 & 25 & 30 \end{pmatrix}.$$

The non-complete distance-regular generalized orbital graphs are

$\Gamma\{2,3\}$,	$\iota = \{22, 11; 1, 2\}$	(and complement)
$\Gamma\{4\}$,	$\iota = \{55, 32; 1, 20\}$	(and complement)
$\Gamma\{2,4\}$,	$\iota = \{66, 35; 1, 30\}$	(and complement)
$\Gamma\{3,4\}$,	$\iota = \{66, 35; 1, 30\}$	(and complement)
$\Gamma\{5\}$,	$\iota = \{66, 35; 1, 30\}$	(and complement)
$\Gamma\{2,4,5\} \cong K_{12\times 12}$		
$\Gamma\{3,4,5\} \cong K_{12\times 12}$.		

The graph $\Gamma\{2,3\}$ is isomorphic to the Hamming graph $H(2,12)$, on which its automorphism group S_{12} wr S_2 acts distance-transitively (see [BCN89, p. 261]).

Each of $\Gamma\{2,4\}$, $\Gamma\{3,4\}$, and $\Gamma\{5\}$ has intersection array $\{66, 35; 1, 30\}$. In each of these graphs, any pair of distinct vertices has exactly $a_1 = 30 = c_2$ common neighbours. Therefore, each of these graphs $\Gamma = (V, E)$ gives rise to a $2 - (144, 66, 30)$ design, whose point-set is V, and block-set is $\{\Gamma(v) \mid v \in V\}$ (see [CvL91, p. 43]). These particular graphs and designs were known to M. Hall [Hal76]. Now $\Gamma\{2,4\} \cong \Gamma\{3,4\}$, as these two graphs are interchanged by an outer automorphism of G. Therefore their corresponding designs are also isomorphic. However, $\Gamma\{2,4\} \not\cong \Gamma\{5\}$. Indeed, using *nauty* (within **GRAPE**), we check that $\text{Aut}(\Gamma\{2,4\}) \cong M_{12}$, but $\text{Aut}(\Gamma\{5\}) \cong M_{12}{:}2$. We also check, using *nauty*, that the designs corresponding to $\Gamma\{2,4\}$ and $\Gamma\{5\}$ are not isomorphic.

(vi) $\pi = \chi_1 + \chi_4 + \chi_5 + \chi_6 + \chi_{11}$, of degree $144 = 1 + 16 + 16 + 45 + 66$.

If G were imprimitive on Ω it would preserve a system of 12 blocks of imprimitivity and one of χ_2, χ_3 would be a constituent of π. Since this is not the case, G is primitive. There is a unique class of maximal subgroups of index 144, $H \cong L_2(11)$, and the associated permutation character is π. For an element $x \in G$ of order 11, $\pi(x) = 1$, and hence all H-orbits in $\Omega \setminus \{\alpha\}$ have length a multiple of 11. An examination of the subgroups of $L_2(11)$ shows that any subgroup of H of index a multiple of, and greater than, 11 must have index at least 55. It follows that H has at least two orbits of length 11 in Ω, and the subdegrees are either 1, 11, 11, 11, 110, or 1, 11, 11, 55, 66. Now H is normalized by an outer automorphism of G, and $H.2 \cong PGL_2(11)$ must permute the H-orbits in Ω. As in case (v), H may not have an odd number of orbits of length 11, and hence the subdegrees are 1, 11, 11, 55, 66.

Thus we have described, up to permutational isomorphism, a unique primitive rank 5 representation with this permutation character. The collapsed adjacency matrices are

$$\begin{pmatrix} 0 & 11 & 0 & 0 & 0 \\ 0 & 0 & 5 & 0 & 6 \\ 1 & 0 & 0 & 10 & 0 \\ 0 & 2 & 0 & 3 & 6 \\ 0 & 0 & 1 & 5 & 5 \end{pmatrix}$$

$$\begin{pmatrix} 0 & 0 & 11 & 0 & 0 \\ 1 & 0 & 0 & 10 & 0 \\ 0 & 5 & 0 & 0 & 6 \\ 0 & 0 & 2 & 3 & 6 \\ 0 & 1 & 0 & 5 & 5 \end{pmatrix}$$

$$\begin{pmatrix} 0 & 0 & 0 & 55 & 0 \\ 0 & 10 & 0 & 15 & 30 \\ 0 & 0 & 10 & 15 & 30 \\ 1 & 3 & 3 & 24 & 24 \\ 0 & 5 & 5 & 20 & 25 \end{pmatrix}$$

$$\begin{pmatrix} 0 & 0 & 0 & 0 & 66 \\ 0 & 0 & 6 & 30 & 30 \\ 0 & 6 & 0 & 30 & 30 \\ 0 & 6 & 6 & 24 & 30 \\ 1 & 5 & 5 & 25 & 30 \end{pmatrix}.$$

The non-complete distance-regular generalized orbital graphs are

$$\Gamma\{5\}, \quad \iota = \{66, 35; 1, 30\} \quad \text{(and complement)}.$$

As in case (v) above, we have that $\Gamma := \Gamma\{5\} = (V, E)$ gives rise to a $2 - (144, 66, 30)$ design, whose point-set is V, and block-set is $\{\Gamma(v) \mid v \in V\}$. Again, this graph and design were known to M. Hall [Hal76]. Using *nauty* (within GRAPE), we check that $\Gamma\{5\}$ has automorphism group $M_{12}:2$, but $\Gamma\{5\}$ is not isomorphic to the distance-regular graph in (v) above having the same intersection array and automorphism group. We also check, using *nauty*, that the design arising from $\Gamma\{5\}$ is not isomorphic to any of the designs with the same parameters discussed in case (v).

(vii) $\pi = \chi_1 + \chi_2 + \chi_7 + \chi_i + \chi_{12}$, where $i = 8$ or 9, or $\pi = \chi_1 + \chi_3 + \chi_7 + \chi_i + \chi_{12}$, where $i = 8$ or 10, of degree $220 = 1 + 11 + 54 + 55 + 99$. (We shall show that, in each case, it is only when $i = 8$ that we get a permutation character.)

By examining the indices of maximal subgroups of G, it is clear that there is no imprimitive permutation representation of G of degree 220. Thus the point stabilizer is a maximal subgroup of G. It follows (from the ATLAS) that $\pi = \chi_1 + \chi_j + \chi_7 + \chi_8 + \chi_{12}$, where $j = 2$ or 3, that these representations correspond to the two classes of maximal subgroups of G having a representative of index 220, and these representations are interchanged by an outer automorphism of G. A point stabilizer is $M_9:S_3$, the stabilizer of an unordered triple of points in the representation with character $\chi_1 + \chi_j$ in (i).

Thus we have described two equivalent, but not permutationally isomorphic, primitive rank 5 representations of M_{12} which are interchanged by an outer automorphism of M_{12}. For each of these representations, the subdegrees are 1, 12, 27, 72, 108, and the collapsed adjacency matrices are

$$\begin{pmatrix} 0 & 12 & 0 & 0 & 0 \\ 1 & 2 & 0 & 0 & 9 \\ 0 & 0 & 0 & 8 & 4 \\ 0 & 0 & 3 & 3 & 6 \\ 0 & 1 & 1 & 4 & 6 \end{pmatrix}$$

$$\begin{pmatrix} 0 & 0 & 27 & 0 & 0 \\ 0 & 0 & 0 & 18 & 9 \\ 1 & 0 & 10 & 0 & 16 \\ 0 & 3 & 0 & 15 & 9 \\ 0 & 1 & 4 & 6 & 16 \end{pmatrix}$$

$$\begin{pmatrix} 0 & 0 & 0 & 72 & 0 \\ 0 & 0 & 18 & 18 & 36 \\ 0 & 8 & 0 & 40 & 24 \\ 1 & 3 & 15 & 14 & 39 \\ 0 & 4 & 6 & 26 & 36 \end{pmatrix}$$

$$\begin{pmatrix} 0 & 0 & 0 & 0 & 108 \\ 0 & 9 & 9 & 36 & 54 \\ 0 & 4 & 16 & 24 & 64 \\ 0 & 6 & 9 & 39 & 54 \\ 1 & 6 & 16 & 36 & 49 \end{pmatrix}.$$

The only non-complete distance-regular generalized orbital graph is

$$\Gamma\{3\} \cong J(12,3), \quad \iota = \{27, 16, 7; 1, 4, 9\}.$$

(*viii*) $\pi = \chi_1 + \chi_6 + \chi_{11} + \chi_{14} + \chi_{15}$, of degree $432 = 1 + 45 + 66 + 144 + 176$.

As there are no maximal subgroups of G of index 432, $H < L < G$, where $m := |G : L| = 12$ or 144, and G preserves a system of imprimitivity $\Sigma := B^G$ where $|\Sigma| = m$, $B := \alpha^L$, and $L = G_B$. If m were 144 then L would be $L_2(11)$ which has no subgroup of index 3. Similarly if m were 12, then L would be M_{11}, which has no subgroup of index 36. Thus π is not a permutation character.

4.4 Automorphism group $M_{12} \colon 2$ of M_{12}

A presentation for $M_{12} \colon 2$ is given in the section for M_{12}.

4.4.1 rank 2

G has no faithful pseudo-permutation character of rank 2.

4.4.2 rank 3

(*i*) $\pi = \chi_1 + \chi_2 + \chi_3$, of degree $24 = 1 + 1 + 22$.

An examination of the table of maximal subgroups of G shows that $H \cong M_{11}$, $H \le M_{12}$. There is just one G-conjugacy class of such subgroups.

As we saw in case (v) of M_{12}, H is transitive by conjugation on the 12 subgroups M_{11} which form one of the two M_{12}-conjugacy classes of such subgroups. Thus this is an imprimitive rank 3 representation, with subdegrees 1, 11, 12, and permutation character π.

The collapsed adjacency matrices are

$$\begin{pmatrix} 0 & 11 & 0 \\ 1 & 10 & 0 \\ 0 & 0 & 11 \end{pmatrix}$$

$$\begin{pmatrix} 0 & 0 & 12 \\ 0 & 0 & 12 \\ 1 & 11 & 0 \end{pmatrix}.$$

4.4.3 rank 4

(ii) $\pi = \chi_1 + \chi_3 + \chi_9 + \chi_{12}$, of degree $144 = 1 + 22 + 55 + 66$.

The restriction of π to M_{12} is the permutation character in case (v) of M_{12}. Thus $H \cong L_2(11).2$, and $H \cap M_{12}$ has orbits on Ω of lengths 1, 11, 11, 55, 66. Since π is the only pseudo-permutation character of G of rank at most 5 whose restriction to M_{12} is the character of case (v) of M_{12} (see the next subsection), it follows that G is primitive of rank 4 on Ω with subdegrees 1, 22, 55, 66, and that there is a unique such representation.

Thus we have described, up to permutational isomorphism, a unique primitive rank 4 representation of $M_{12}{:}2$. The collapsed adjacency matrices are

$$\begin{pmatrix} 0 & 22 & 0 & 0 \\ 1 & 10 & 5 & 6 \\ 0 & 2 & 8 & 12 \\ 0 & 2 & 10 & 10 \end{pmatrix}$$

$$\begin{pmatrix} 0 & 0 & 55 & 0 \\ 0 & 5 & 20 & 30 \\ 1 & 8 & 22 & 24 \\ 0 & 10 & 20 & 25 \end{pmatrix}$$

$$\begin{pmatrix} 0 & 0 & 0 & 66 \\ 0 & 6 & 30 & 30 \\ 0 & 12 & 24 & 30 \\ 1 & 10 & 25 & 30 \end{pmatrix}.$$

The non-complete distance-regular generalized orbital graphs are

$$\Gamma\{2\}, \quad \iota = \{22, 11; 1, 2\} \quad \text{(and complement)}$$
$$\Gamma\{3\}, \quad \iota = \{55, 32; 1, 20\} \quad \text{(and complement)}$$
$$\Gamma\{4\}, \quad \iota = \{66, 35; 1, 30\} \quad \text{(and complement)}.$$

(iii) $\pi = \chi_1 + \chi_4 + \chi_5 + \chi_{12}$, of degree $144 = 1 + 32 + 45 + 66$.

The restriction of π to M_{12} is the permutation character in case (vi) of M_{12}. Thus $H \cong L_2(11).2$, and $H \cap M_{12}$ has orbits on Ω of lengths 1, 11, 11, 55, 66. As in the previous case, since π is the only pseudo-permutation character of G of rank at most 5 whose restriction to M_{12} is the character of case (vi) of M_{12}, it follows that G is primitive of rank 4 on Ω with subdegrees 1, 22, 55, 66, and that there is a unique such representation.

Thus we have described, up to permutational isomorphism, a unique primitive rank 4 representation of $M_{12}{:}\,2$. The collapsed adjacency matrices are

$$\begin{pmatrix} 0 & 22 & 0 & 0 \\ 1 & 5 & 10 & 6 \\ 0 & 4 & 6 & 12 \\ 0 & 2 & 10 & 10 \end{pmatrix}$$

$$\begin{pmatrix} 0 & 0 & 55 & 0 \\ 0 & 10 & 15 & 30 \\ 1 & 6 & 24 & 24 \\ 0 & 10 & 20 & 25 \end{pmatrix}$$

$$\begin{pmatrix} 0 & 0 & 0 & 66 \\ 0 & 6 & 30 & 30 \\ 0 & 12 & 24 & 30 \\ 1 & 10 & 25 & 30 \end{pmatrix}.$$

The non-complete distance-regular generalized orbital graphs are

$$\Gamma\{4\}, \quad \iota = \{66, 35; 1, 30\} \quad \text{(and complement)}.$$

4.4.4 rank 5

(iv) $\pi = \chi_1 + \chi_2 + \chi_3 + \chi_7 + \chi_8$, of degree $132 = 1 + 1 + 22 + 54 + 54$.

The restriction of π to M_{12} is the permutation character $2\chi_1 + \chi_2 + \chi_3 + 2\chi_7$. This means in particular that M_{12} has two orbits, say Ω_1 and Ω_2, on Ω, each of length 66, and the permutation characters for M_{12} on Ω_1 and Ω_2 must be the characters in case (ii) of M_{12}.

We may suppose that H is the stabilizer of a point in Ω_1. Thus $H \cong M_{10}{:}2$, $H < M_{12}$, and H has orbits of lengths 1, 20, 45 on Ω_1. Further, the number of orbits of H on Ω_2 is equal to the inner product of the permutation characters of M_{12} on Ω_1 and Ω_2, namely 2. Thus G has an imprimitive rank 5 representation on Ω. The subgroup M_{10} of index 2 in H is the stabilizer of two points in one of the representations of M_{12} of degree 12, say on a set Δ_1. Its action on Ω_2 is permutationally isomorphic to its action on unordered pairs from the other representation of M_{12} of degree 12, say on Δ_2. Since M_{10} has index 11 in the stabilizer L of a point of Δ_1, and since L is transitive on Δ_2, it follows that each orbit of H on Ω_2 has length divisible by 6. Further, since an element of order 5 in H fixes exactly one point of Ω_2, the H-orbit lengths on Ω_2 must be either 6, 60 or 30, 36. However, since H has no transitive representation of degree 6, it follows that the subdegrees of G are 1, 20, 30, 36, 45.

Thus we have described, up to permutational isomorphism, a unique imprimitive rank 5 representation of $M_{12}{:}2$. The collapsed adjacency matrices are

$$\begin{pmatrix} 0 & 20 & 0 & 0 & 0 \\ 1 & 10 & 0 & 0 & 9 \\ 0 & 0 & 8 & 12 & 0 \\ 0 & 0 & 10 & 10 & 0 \\ 0 & 4 & 0 & 0 & 16 \end{pmatrix}$$

$$\begin{pmatrix} 0 & 0 & 30 & 0 & 0 \\ 0 & 0 & 12 & 18 & 0 \\ 1 & 8 & 0 & 0 & 21 \\ 0 & 10 & 0 & 0 & 20 \\ 0 & 0 & 14 & 16 & 0 \end{pmatrix}$$

$$\begin{pmatrix} 0 & 0 & 0 & 36 & 0 \\ 0 & 0 & 18 & 18 & 0 \\ 0 & 12 & 0 & 0 & 24 \\ 1 & 10 & 0 & 0 & 25 \\ 0 & 0 & 16 & 20 & 0 \end{pmatrix}$$

$$\begin{pmatrix} 0 & 0 & 0 & 0 & 45 \\ 0 & 9 & 0 & 0 & 36 \\ 0 & 0 & 21 & 24 & 0 \\ 0 & 0 & 20 & 25 & 0 \\ 1 & 16 & 0 & 0 & 28 \end{pmatrix}.$$

The only non-complete distance-regular generalized orbital graph is

$$\Gamma\{3,4\} \cong K_{2\times 66}.$$

(v) $\pi = \chi_1 + \chi_5 + \chi_{12} + \chi_{18} + \chi_{21}$, of degree $432 = 1 + 45 + 66 + 144 + 176$.

The restriction of π to M_{12} is the pseudo-permutation character of case (viii) of M_{12}. Since that character is not a permutation character of M_{12}, it follows that π is not a permutation character of $M_{12}\!:\!2$.

4.5 The Janko group J_1

A presentation for J_1 is

$$\langle a3b5c3d5e \mid 1 = (abc)^5 [= (bcd)^5], a = (cde)^5 \rangle,$$

in which

$$L_2(11) \cong \langle a, b, c, d \rangle.$$

This presentation was discovered by J.H. Conway and R.A. Parker (see [Soi87b]).

4.5.1 rank 2

(i) $\pi = \chi_1 + \chi_4$ of degree $77 = 1 + 76$.

G has no subgroup of index 77.

4.5.2 rank 3

(*ii*) $\pi = \chi_1 + \chi_4 + \chi_6$ of degree $154 = 1 + 76 + 77$.

G has no subgroup of index 154.

4.5.3 rank 4

G has no pseudo-permutation character of rank 4.

4.5.4 rank 5

(*iii*) $\pi = \chi_1 + \chi_2 + \chi_3 + \chi_4 + \chi_6$ of degree $266 = 1 + 56 + 56 + 76 + 77$.

There is a unique class of subgroups of G of index 266, with associated permutation character π. A point stabilizer is a maximal $H \cong L_2(11)$.

The subdegrees are 1, 11, 12, 110, 132, and the collapsed adjacency matrices are

$$\begin{pmatrix} 0 & 11 & 0 & 0 & 0 \\ 1 & 0 & 0 & 10 & 0 \\ 0 & 0 & 0 & 0 & 11 \\ 0 & 1 & 0 & 4 & 6 \\ 0 & 0 & 1 & 5 & 5 \end{pmatrix}$$

$$\begin{pmatrix} 0 & 0 & 12 & 0 & 0 \\ 0 & 0 & 0 & 0 & 12 \\ 1 & 0 & 0 & 0 & 11 \\ 0 & 0 & 0 & 6 & 6 \\ 0 & 1 & 1 & 5 & 5 \end{pmatrix}$$

$$\begin{pmatrix} 0 & 0 & 0 & 110 & 0 \\ 0 & 10 & 0 & 40 & 60 \\ 0 & 0 & 0 & 55 & 55 \\ 1 & 4 & 6 & 45 & 54 \\ 0 & 5 & 5 & 45 & 55 \end{pmatrix}$$

$$\begin{pmatrix} 0 & 0 & 0 & 0 & 132 \\ 0 & 0 & 12 & 60 & 60 \\ 0 & 11 & 11 & 55 & 55 \\ 0 & 6 & 6 & 54 & 66 \\ 1 & 5 & 5 & 55 & 66 \end{pmatrix}.$$

The only non-complete distance-regular generalized orbital graph is

$$\Gamma\{2\}, \quad \iota = \{11, 10, 6, 1; 1, 1, 5, 11\}, \quad G \text{ acts distance-transitively.}$$

This graph is known as the Livingstone graph, and is described in detail in [BCN89].

4.6 The Mathieu group M_{22}

A presentation for M_{22} is

$$\langle a3b5c3d, a4e3c \mid 1 = (abc)^5 [= (bcd)^5],$$

$$1 = (eab)^3 = (bce)^5 = (aecd)^4 = (abce)^8 \rangle$$

(see [Soi85, Soi88]), in which

$$M_{21} \cong L_3(4) \cong \langle a, b, c, a^{bcde} \rangle,$$

$$2^4 : A_6 \cong \langle a, b, c, e \rangle > \langle b, c, e, (aec)^3 \rangle \cong 2^4 : L_2(5),$$

$$A_7 \cong \langle b, c, d, d^{cbae} \rangle,$$

$$2^4 : S_5 \cong \langle a, c, d, e, bcdecb \rangle,$$

$$2^3 : L_3(2) \cong \langle a, c, d, e, a^{ecb} \rangle,$$

and

$$M_{10} \cong \langle b, c, e, abcdcbeacbcdcbaecbacd \rangle.$$

The short words generating $2^4 : S_5$, and those generating $2^3 : L_3(2)$, were found by M. Schönert.

We remark that the subgroup $L_2(11) \cong \langle a, b, c, d \rangle$ leads to a rank 6 representation of M_{22}, and a vertex-transitive distance-regular graph of diameter 3 (see [Soi95]).

A presentation for $M_{22}\!:\!2$ is obtained from that of M_{22}, above, by adjoining a generator t, and the relations

$$1 = t^2 = (at)^2 = (ct)^2 = (dt)^2 = (et)^2 = b^t be.$$

We note that t normalizes each of the following of the specific subgroups of M_{22} given above: $\langle a, b, c, e \rangle$, $\langle b, c, e, (aec)^3 \rangle$, $\langle a, c, d, e, bcdecb \rangle$, $\langle a, c, d, e, a^{ecb} \rangle$, and $\langle b, c, e, abcdcbeacbcdcbaecbacd \rangle$. The element t does not normalize the specific subgroup $\langle a, b, c, a^{bcde} \rangle \cong M_{21}$, but the (outer) element $eabt$ does.

4.6.1 rank 2

(i) $\pi = \chi_1 + \chi_2$, of degree $22 = 1 + 21$.

This character corresponds to the unique 2-transitive representation of G of degree 22. A point stabilizer is $H \cong M_{21} \cong L_3(4)$.

In the remainder of this section, let X be the set of points of size 22 in this representation. Then G preserves a Steiner system $S(3, 6, 22)$ with point-set X and block-set $\mathcal{B}$ consisting of 77 subsets of X of size 6, called *hexads*.

(ii) $\pi = \chi_1 + \chi_5$, of degree $56 = 1 + 55$.

G has no subgroup of index 56.

4.6.2 rank 3

(iii) $\pi = \chi_1 + \chi_2 + \chi_5$, of degree $77 = 1 + 21 + 55$.

By examining the indices of maximal subgroups of G, it is clear that there is a unique class of subgroups of G of index 77. The associated permutation representation is the primitive rank 3 representation of G on $\mathcal{B}$, with character π. A point stabilizer is $H \cong 2^4\!:\!A_6$.

The subdegrees are 1, 16, 60, and the collapsed adjacency matrices are

$$\begin{pmatrix} 0 & 16 & 0 \\ 1 & 0 & 15 \\ 0 & 4 & 12 \end{pmatrix}$$

$$\begin{pmatrix} 0 & 0 & 60 \\ 0 & 15 & 45 \\ 1 & 12 & 47 \end{pmatrix}.$$

(iv) $\pi = \chi_1 + \chi_2 + \chi_7$, of degree $176 = 1 + 21 + 154$.

There are exactly two classes of maximal subgroups of G of index 176, with associated permutation character π, each having a representative isomorphic to A_7. The associated representations are interchanged by an outer automorphism of G.

For each of these two equivalent, but not permutationally isomorphic, representations, the subdegrees are 1, 70, 105, and the collapsed adjacency matrices are

$$\begin{pmatrix} 0 & 70 & 0 \\ 1 & 18 & 51 \\ 0 & 34 & 36 \end{pmatrix}$$

$$\begin{pmatrix} 0 & 0 & 105 \\ 0 & 51 & 54 \\ 1 & 36 & 68 \end{pmatrix}.$$

If there were also an imprimitive permutation representation of G with character π, then we would have $H < L < G$ with $|G : L|$ a proper divisor of 176. Then $L \cong L_3(4)$ of index 22, but this subgroup L has no subgroup of index 8.

(v) $\pi = \chi_1 + \chi_5 + \chi_7$, of degree $210 = 1 + 55 + 154$.

G has no subgroup of index 210.

4.6.3   rank 4

(vi) $\pi = \chi_1 + \chi_2 + \chi_5 + \chi_6$, of degree $176 = 1 + 21 + 55 + 99$.

Arguing as in *(iv)* above, the only subgroups of G of index 176 are isomorphic to A_7, but then the associated permutation representation has rank 3. Thus π is not a permutation character.

(vii) $\pi = \chi_1 + \chi_2 + \chi_5 + \chi_7$, of degree $231 = 1 + 21 + 55 + 154$.

There is a unique class of maximal subgroups of G of index 231 with associated permutation character π. A point stabilizer is $H \cong 2^4{:}S_5$, the stabilizer of an unordered pair of points in the representation with character $\chi_1 + \chi_2$ in (i).

The subdegrees are 1, 30, 40, 160, and the collapsed adjacency matrices are

$$\begin{pmatrix} 0 & 30 & 0 & 0 \\ 1 & 9 & 4 & 16 \\ 0 & 3 & 3 & 24 \\ 0 & 3 & 6 & 21 \end{pmatrix}$$

$$\begin{pmatrix} 0 & 0 & 40 & 0 \\ 0 & 4 & 4 & 32 \\ 1 & 3 & 20 & 16 \\ 0 & 6 & 4 & 30 \end{pmatrix}$$

$$\begin{pmatrix} 0 & 0 & 0 & 160 \\ 0 & 16 & 32 & 112 \\ 0 & 24 & 16 & 120 \\ 1 & 21 & 30 & 108 \end{pmatrix}.$$

The non-complete distance-regular generalized orbital graphs are

$$\Gamma\{2\}, \qquad \iota = \{30, 20; 1, 3\} \qquad \text{(and complement)}$$
$$\Gamma\{3\} \cong J(22, 2), \qquad \iota = \{40, 19; 1, 4\} \qquad \text{(and complement)}.$$

The graph $\Gamma\{2\}$ was originally discovered by P.J. Cameron (see also [IKF84, FKM94]).

If there were also an imprimitive permutation representation of G with character π, then we would have $H < L < G$ with $|G : L|$ a proper divisor of 231. Then $L \cong 2^4 : A_6$ of index 77, but this subgroup L has no subgroup of index 3.

$(viii)$ $\quad \pi = \chi_1 + \chi_2 + \chi_5 + \chi_9$, of degree $308 = 1 + 21 + 55 + 231$.

G has no maximal subgroup of index 308, and hence $H < L < G$, where $m := |G : L|$ is either 22 or 77, and G preserves a system of imprimitivity $\Sigma := B^G$ where $|\Sigma| = m$, $B := \alpha^L$, and $L = G_B$. If $m = 77$, then $L \cong 2^4 : A_6$, which has no subgroup of index 4. Similarly, if $m = 22$, then $L \cong L_3(4)$, which has no subgroup of index 14. Thus π is not a permutation character.

4.6.4 rank 5

(ix) $\pi = \chi_1 + \chi_2 + \chi_5 + \chi_6 + \chi_7$, of degree $330 = 1 + 21 + 55 + 99 + 154$.

There is a unique class of maximal subgroups with associated permutation character π. A point stabilizer is $H \cong 2^3 : L_3(2)$.

The subdegrees are 1, 7, 42, 112, 168, and the collapsed adjacency matrices are

$$\begin{pmatrix} 0 & 7 & 0 & 0 & 0 \\ 1 & 0 & 6 & 0 & 0 \\ 0 & 1 & 2 & 0 & 4 \\ 0 & 0 & 0 & 1 & 6 \\ 0 & 0 & 1 & 4 & 2 \end{pmatrix}$$

$$\begin{pmatrix} 0 & 0 & 42 & 0 & 0 \\ 0 & 6 & 12 & 0 & 24 \\ 1 & 2 & 7 & 16 & 16 \\ 0 & 0 & 6 & 18 & 18 \\ 0 & 1 & 4 & 12 & 25 \end{pmatrix}$$

$$\begin{pmatrix} 0 & 0 & 0 & 112 & 0 \\ 0 & 0 & 0 & 16 & 96 \\ 0 & 0 & 16 & 48 & 48 \\ 1 & 1 & 18 & 44 & 48 \\ 0 & 4 & 12 & 32 & 64 \end{pmatrix}$$

$$\begin{pmatrix} 0 & 0 & 0 & 0 & 168 \\ 0 & 0 & 24 & 96 & 48 \\ 0 & 4 & 16 & 48 & 100 \\ 0 & 6 & 18 & 48 & 96 \\ 1 & 2 & 25 & 64 & 76 \end{pmatrix}.$$

The only non-complete distance-regular generalized orbital graph is

$\Gamma\{2\}$, $\iota = \{7, 6, 4, 4; 1, 1, 1, 6\}$, G acts distance-transitively.

We may take the vertices of $\Gamma\{2\}$ to be the 330 octads of an $S(5, 8, 24)$ which do not contain either of two given points. We join two such octads by an edge in $\Gamma\{2\}$ precisely when they are disjoint. This graph is described in more detail in [BCN89].

If there were also an imprimitive permutation representation with character π, then we would have $H < L < G$ with $m := |G : L|$ dividing

330. Then $m = 22$ and $L \cong L_3(4)$. However this subgroup L has no subgroup of index 15.

(x) $\pi = \chi_1 + \chi_2 + \chi_5 + \chi_7 + \chi_9$, of degree $462 = 1 + 21 + 55 + 154 + 231$.

G has no maximal subgroup of index 462, so $H < L < G$ with $m :=$ $|G : L|$ dividing 462, and G preserves a system of imprimitivity $\Sigma := B^G$ where $|\Sigma| = m$, $B := \alpha^L$, and $L = G_B$. We observe that m is 22, 77, or 231, and L is $L_3(4)$, $2^4 : A_6$, or $2^4 : S_5$ respectively.

Suppose first that $m = 22$. Then L is the stabilizer of a point x of X. We may identify $X \setminus \{x\}$ with the set of 21 points of the projective plane $PG_2(4)$. By Lemma 3.4, H is not the stabilizer in L of a point of $X \setminus \{x\}$, and hence H is the stabilizer of a hyperplane of $PG_2(4)$, and has orbits on X of lengths 1, 5, 16. The actions of G on X and Σ are permutationally isomorphic, and so H has orbits on Σ of lengths 1, 5, 16, and hence fixes setwise subsets of Ω of sizes 21, 21.5, and 21.16 respectively. Since $|H|$ is not divisible by 7, H has at least two orbits in each of these three sets, so G has rank at least 6, which is a contradiction.

Next let $m = 77$. Then H has index 6 in $L \cong 2^4 : A_6$. In particular, L is 2-transitive on B, so H has orbits of lengths 1, 5 in B. Further, the action of G on Σ is permutationally isomorphic to its rank 3 action on the set $\mathcal{B}$ of hexads. Hence L has orbits on Σ of lengths 1, 16, 60, and L is the stabilizer of a hexad $Y \in \mathcal{B}$. Moreover, either the actions of L on B and Y are permutationally isomorphic, or the stabilizer H of the point α of B induces a 2-transitive action on Y with $H^Y \cong L_2(5)$. In the former case, the action of G on Ω is permutationally isomorphic to its action on incident point-hexad pairs from the Steiner system, and so H is also a subgroup of index 21 in the stabilizer of a point of X. We showed in the previous paragraph that this is not possible. Thus $H^Y \cong L_2(5)$. Direct calculation shows that for this $H \cong 2^4 : L_2(5)$, the action of G is rank 5, with subdegrees 1, 5, 96, 120, 240, and collapsed adjacency matrices

$$\begin{pmatrix} 0 & 5 & 0 & 0 & 0 \\ 1 & 4 & 0 & 0 & 0 \\ 0 & 0 & 5 & 0 & 0 \\ 0 & 0 & 0 & 1 & 4 \\ 0 & 0 & 0 & 2 & 3 \end{pmatrix}$$

$$\begin{pmatrix} 0 & 0 & 96 & 0 & 0 \\ 0 & 0 & 96 & 0 & 0 \\ 1 & 5 & 0 & 30 & 60 \\ 0 & 0 & 24 & 24 & 48 \\ 0 & 0 & 24 & 24 & 48 \end{pmatrix}$$

$$\begin{pmatrix} 0 & 0 & 0 & 120 & 0 \\ 0 & 0 & 0 & 24 & 96 \\ 0 & 0 & 30 & 30 & 60 \\ 1 & 1 & 24 & 34 & 60 \\ 0 & 2 & 24 & 30 & 64 \end{pmatrix}$$

$$\begin{pmatrix} 0 & 0 & 0 & 0 & 240 \\ 0 & 0 & 0 & 96 & 144 \\ 0 & 0 & 60 & 60 & 120 \\ 0 & 4 & 48 & 60 & 128 \\ 1 & 3 & 48 & 64 & 124 \end{pmatrix}.$$

The only non-complete distance-regular generalized orbital graph is

$$\Gamma\{3,4,5\} \cong K_{77 \times 6}.$$

Now if $m = 231$, then H has index 2 in $L \cong 2^4 : S_5$. Thus $H \cong 2^4 : A_5$, which is the stabilizer of an ordered pair of distinct points of X. By Lemma 3.4, G has rank at least 7 on the cosets of H.

(xi) $\pi = \chi_1 + \chi_2 + \chi_5 + \chi_7 + \chi_{12}$, of degree $616 = 1 + 21 + 55 + 154 + 385$.

There is a unique class of maximal subgroups of G of index 616 with associated character π. The point stabilizer is $H \cong M_{10} \cong A_6 \cdot 2_3$.

The subdegrees are 1, 30, 45, 180, 360, and the collapsed adjacency matrices are

$$\begin{pmatrix} 0 & 30 & 0 & 0 & 0 \\ 1 & 14 & 3 & 0 & 12 \\ 0 & 2 & 4 & 0 & 24 \\ 0 & 0 & 0 & 14 & 16 \\ 0 & 1 & 3 & 8 & 18 \end{pmatrix}$$

$$\begin{pmatrix} 0 & 0 & 45 & 0 & 0 \\ 0 & 3 & 6 & 0 & 36 \\ 1 & 4 & 0 & 32 & 8 \\ 0 & 0 & 8 & 5 & 32 \\ 0 & 3 & 1 & 16 & 25 \end{pmatrix}$$

$$\begin{pmatrix} 0 & 0 & 0 & 180 & 0 \\ 0 & 0 & 0 & 84 & 96 \\ 0 & 0 & 32 & 20 & 128 \\ 1 & 14 & 5 & 72 & 88 \\ 0 & 8 & 16 & 44 & 112 \end{pmatrix}$$

$$\begin{pmatrix} 0 & 0 & 0 & 0 & 360 \\ 0 & 12 & 36 & 96 & 216 \\ 0 & 24 & 8 & 128 & 200 \\ 0 & 16 & 32 & 88 & 224 \\ 1 & 18 & 25 & 112 & 204 \end{pmatrix}.$$

There are no non-complete distance-regular generalized orbital graphs.

If there were also an imprimitive permutation representation with this character, then we would have $H < L < G$ with $m := |G : L|$ dividing 616. Then m is 22 or 77, and H is a subgroup of index 28 in $L \cong L_3(4)$ or index 8 in $L \cong 2^4 : A_6$, respectively. In neither case does such a subgroup exist.

(*xii*) $\pi = \chi_1 + \chi_2 + \chi_5 + \chi_9 + \chi_{12}$, of degree $693 = 1 + 21 + 55 + 231 + 385$.

There is no maximal subgroup of G of index 693, and so $H < L < G$ with $m := |G : L|$ dividing 693. Then m is 77 or 231, and H is a subgroup of index 9 in $2^4 : A_6$ or index 3 in $2^4 : S_5$, respectively. In neither case does such a subgroup exist.

(*xiii*) $\pi = \chi_1 + \chi_2 + \chi_7 + \chi_9 + \chi_{12}$, of degree $792 = 1 + 21 + 154 + 231 + 385$.

There is no maximal subgroup of G of index 792, and so $H < L < G$ with $m := |G : L|$ dividing 792. Now m must be 22, and H is a subgroup of index 36 in $L_3(4)$. However there is no such subgroup.

(*xiv*) $\pi = \chi_1 + \chi_3 + \chi_4 + \chi_7 + \chi_{12}$, of degree $630 = 1 + 45 + 45 + 154 + 385$.

G has no subgroup of index 630.

4.7 Automorphism group $M_{22}:2$ of M_{22}

A presentation for $M_{22}:2$ is given in the section for M_{22}.

4.7.1 rank 2

There is just one faithful pseudo-permutation character of G of rank 2.

(i) $\pi = \chi_1 + \chi_3$, of degree $22 = 1 + 21$.

This is the character of the unique 2-transitive representation of G of degree 22. A point stabilizer is $H \cong L_3(4):2_2$, and the restriction of the representation to G has the permutation character of case (i) for M_{22}.

4.7.2 rank 3

(ii) $\pi = \chi_1 + \chi_3 + \chi_9$, of degree $77 = 1 + 21 + 55$.

There is a unique class of subgroups of G of index 77. The associated permutation representation is the primitive rank 3 representation of G on $\mathcal{B}$, with character π. A point stabilizer is $H \cong 2^4:S_6$, and the restriction to M_{22} is the representation in case (iii) for M_{22}. The collapsed adjacency matrices are the same as for that representation.

4.7.3 rank 4

(iii) $\pi = \chi_1 + \chi_2 + \chi_3 + \chi_4$, of degree $44 = 1 + 1 + 21 + 21$.

There is no maximal subgroup of G of index 44, and so G is imprimitive, and $H \cong M_{21} < M_{22} < G$. The restriction of π to M_{22} is intransitive with two orbits on each of which the action of M_{22} is permutationally isomorphic to the 2-transitive action in case (i) of M_{22}. The subdegrees are therefore 1, 1, 21, 21, and the collapsed adjacency matrices are

$$\begin{pmatrix} 0 & 1 & 0 & 0 \\ 1 & 0 & 0 & 0 \\ 0 & 0 & 0 & 1 \\ 0 & 0 & 1 & 0 \end{pmatrix}$$

$$\begin{pmatrix} 0 & 0 & 21 & 0 \\ 0 & 0 & 0 & 21 \\ 1 & 0 & 20 & 0 \\ 0 & 1 & 0 & 20 \end{pmatrix}$$

$$\begin{pmatrix} 0 & 0 & 0 & 21 \\ 0 & 0 & 21 & 0 \\ 0 & 1 & 0 & 20 \\ 1 & 0 & 20 & 0 \end{pmatrix}.$$

The non-complete distance-regular generalized orbital graphs are

$\Gamma\{4\}$, $\iota = \{21, 20, 1; 1, 20, 21\}$, G acts distance-transitively
$\Gamma\{2, 4\} \cong K_{2 \times 22}$
$\Gamma\{3, 4\} \cong K_{22 \times 2}$.

The graph $\Gamma\{4\}$ is an example of a Taylor graph. In fact, $\Gamma\{4\}$ is just the graph $K_{2 \times 22}$ with a 1-factor removed, and has automorphism group $2 \times S_{22}$.

(iv) $\pi = \chi_1 + \chi_2 + \chi_9 + \chi_{10}$, of degree $112 = 1 + 1 + 55 + 55$.

G has no subgroup of index 112.

(v) $\pi = \chi_1 + \chi_3 + \chi_9 + \chi_{13}$, of degree $231 = 1 + 21 + 55 + 154$.

There is a unique class of maximal subgroups of index 231 in G, and it corresponds to the primitive permutation representation of G with character π. A point stabilizer is $H \cong 2^5 : S_5$, and the restriction to M_{22} is the primitive rank 4 representation of case (vii) of M_{22}. Thus the collapsed adjacency matrices and associated distance-regular graphs are the same as for that representation.

If there were also an imprimitive representation with character π, then we would have $H < L < G$ with H of index 3 in $L \cong 2^4 : S_6$. No such subgroup exists.

4.7.4 rank 5

(vi) $\pi = \chi_1 + \chi_3 + \chi_9 + \chi_{11} + \chi_{13}$, of degree $330 = 1 + 21 + 55 + 99 + 154$.

There is a unique primitive permutation representation of G of degree 330, and it has associated character π. A point stabilizer is $H \cong$

$2^3\colon L_3(2) \times 2$, and the restriction to M_{22} is the primitive rank 5 representation of case (ix) of M_{22}. Thus the collapsed adjacency matrices and the associated non-complete distance-regular graph, on which G acts distance-transitively, are the same as for that representation.

If there were also an imprimitive permutation representation with character π, then we would have $H < L < G$ with $m := |G : L|$ dividing 330. Then $m = 22$ and $L \cong L_3(4)\colon 2_2$. However this subgroup L has no subgroup of index 15.

(vii) $\pi = \chi_1 + \chi_3 + \chi_9 + \chi_{13} + \chi_{14}$, of degree $385 = 1 + 21 + 55 + 154 + 154$.

There is no maximal subgroup of G of index 385. Indeed, we would need to have $H < L < G$, with H of index 5 in $L \cong 2^4\colon S_6$, but no such subgroup exists.

$(viii)$ $\pi = \chi_1 + \chi_3 + \chi_9 + \chi_{13} + \chi_{18}$, of degree $462 = 1 + 21 + 55 + 154 + 231$.

The restriction of π to M_{22} is the rank 5 permutation character of case (x) for M_{22}. There is a unique transitive permutation representation of M_{22} with this restricted character. This representation is imprimitive, preserving a set Σ of 77 blocks of size 6, and has point stabilizer $H \cap M_{22} \cong 2^4\colon L_2(5) < 2^4\colon A_6$. It follows that $H \cong 2^4\colon PGL_2(5) < 2^4\colon S_6$, there is a unique representation of G with character π, and the collapsed adjacency matrices and associated non-complete distance-regular graph are the same as in case (x) for M_{22}.

(ix) $\pi = \chi_1 + \chi_3 + \chi_9 + \chi_{13} + \chi_{20}$, of degree $616 = 1 + 21 + 55 + 154 + 385$.

There is a unique class of maximal subgroups of G of index 616 having associated permutation character π. A point stabilizer is $H \cong A_6 \cdot 2^2$, and the restriction to M_{22} is the primitive rank 5 representation of case (xi) of M_{22}. Thus the collapsed adjacency matrices are the same as for that representation, and there is no associated non-complete distance-regular graph.

If there were also an imprimitive representation with character π, then we would have $H < L < G$ with $m = 22$ or 77. In the former case, H has index 28 in $L \cong L_3(4)\colon 2_2$, but L has no such subgroup. In the latter case, H has index 8 in $L \cong 2^4\colon S_6$. Again L has no such subgroup.

4.8 The Hall-Janko group J_2

A presentation for J_2 is

$$\langle a5b3c8d3e \mid a = (cd)^4, 1 = ((bcdcd)^5 e)^3 [= (abc)^5]\rangle$$

(see [Soi85]), in which

$$3 \cdot PGL_2(9) \cong \langle a, b, c, d\rangle.$$

The representation of J_2 on the cosets of $3 \cdot PGL_2(9)$ extends to a representation of the same rank for $J_2: 2$ on the cosets of $3 \cdot A_6 \cdot 2^2$.

A presentation for $J_2: 2$ is

$$\langle a3b3c8d3e3f \mid a = (cd)^4, 1 = (bcde)^8\rangle$$

(see [Soi85]), in which

$$U_3(3): 2 \cong \langle a, b, c, d, e\rangle,$$

and

$$2^{1+4}_- . S_5 \cong \langle a, c, d, e, f, e^{dcbadcb}\rangle.$$

The representation of $J_2: 2$ on the cosets of $U_3(3): 2$ restricts to a representation of the same rank for J_2 on the cosets of $U_3(3)$.

4.8.1 rank 2

G has no pseudo-permutation character of rank 2.

4.8.2 rank 3

(i) $\quad \pi = \chi_1 + \chi_6 + \chi_7$ of degree $100 = 1 + 36 + 63$.

There is a unique class of subgroups of G of index 100. A representative is a maximal $H \cong U_3(3)$, and the associated permutation character is π.

The subdegrees are 1, 36, 63, and the collapsed adjacency matrices are

$$\begin{pmatrix} 0 & 36 & 0 \\ 1 & 14 & 21 \\ 0 & 12 & 24 \end{pmatrix}$$

$$\begin{pmatrix} 0 & 0 & 63 \\ 0 & 21 & 42 \\ 1 & 24 & 38 \end{pmatrix}.$$

4.8.3 rank 4

(ii) $\pi = \chi_1 + \chi_7 + \chi_{10} + \chi_{11}$ of degree $280 = 1 + 63 + 90 + 126$.

There is a unique class of subgroups of G of index 280. A representative is a maximal $H \cong 3 \cdot PGL_2(9)$, and the associated permutation character is π.

The subdegrees are 1, 36, 108, 135, and the collapsed adjacency matrices are

$$\begin{pmatrix} 0 & 36 & 0 & 0 \\ 1 & 8 & 12 & 15 \\ 0 & 4 & 12 & 20 \\ 0 & 4 & 16 & 16 \end{pmatrix}$$

$$\begin{pmatrix} 0 & 0 & 108 & 0 \\ 0 & 12 & 36 & 60 \\ 1 & 12 & 40 & 55 \\ 0 & 16 & 44 & 48 \end{pmatrix}$$

$$\begin{pmatrix} 0 & 0 & 0 & 135 \\ 0 & 15 & 60 & 60 \\ 0 & 20 & 55 & 60 \\ 1 & 16 & 48 & 70 \end{pmatrix}.$$

The non-complete distance-regular generalized orbital graphs are

$$\Gamma\{2\}, \quad \iota = \{36, 27; 1, 4\} \quad \text{(and complement)}$$
$$\Gamma\{4\}, \quad \iota = \{135, 64; 1, 60\} \quad \text{(and complement)}.$$

These distance-regular graphs were originally presented in [IKF84] (see also [FKM94]).

We may take the vertex-sets of $\Gamma\{2\}$ and $\Gamma\{4\}$ to be the conjugacy class in J_2 of $3A$-generated subgroups of order 3. Then vertices A, B are joined by an edge in $\Gamma\{2\}$ (respectively $\Gamma\{4\}$) if and only if $\langle A, B \rangle$ is isomorphic to A_4 (respectively $SL_2(3)$).

4.8.4 rank 5

(iii) $\pi = \chi_1 + \chi_6 + \chi_7 + \chi_{10} + \chi_{12}$ of degree $350 = 1 + 36 + 63 + 90 + 160$.
G has no subgroup of index 350.

4.9 Automorphism group $J_2{:}2$ of J_2

A presentation for $J_2{:}2$ is given in the section for J_2.

4.9.1 rank 2

G has no faithful pseudo-permutation character of rank 2.

4.9.2 rank 3

(i) $\pi = \chi_1 + \chi_5 + \chi_7$ of degree $100 = 1 + 36 + 63$.

There is a unique class of subgroups of G of index 100. A representative is a maximal $H \cong U_3(3){:}2$, and the associated permutation character is π. The restriction of this representation to J_2 is the representation in case (i) for J_2, and so the collapsed adjacency matrices are the same as for that representation.

4.9.3 rank 4

(ii) $\pi = \chi_1 + \chi_2 + \chi_3 + \chi_4$ of degree $72 = 1 + 1 + 28 + 42$.

G has no subgroup of index 72.

(iii) $\pi = \chi_1 + \chi_7 + \chi_{10} + \chi_{12}$ of degree $280 = 1 + 63 + 90 + 126$.

There is a unique class of subgroups of G of index 280. A representative is a maximal $H \cong 3{\cdot}A_6{\cdot}2^2$, and the associated permutation character is π. The restriction of this representation to J_2 is the representation in case (ii) for J_2, and so the collapsed adjacency matrices and associated distance-regular graphs are the same as for that representation.

4.9.4 rank 5

(iv) $\pi = \chi_1 + \chi_2 + \chi_3 + \chi_{10} + \chi_{11}$ of degree $210 = 1 + 1 + 28 + 90 + 90$.
G has no subgroup of index 210.

(v) $\pi = \chi_1 + \chi_3 + \chi_5 + \chi_{10} + \chi_{14}$ of degree $315 = 1 + 28 + 36 + 90 + 160$.
There is a unique class of subgroups of G of index 315. A representative is a maximal $H \cong 2_-^{1+4}.S_5$, and the associated permutation character is π.

The subdegrees are 1, 10, 64, 80, 160. (The restriction of this representation to J_2 has rank 6 and subdegrees 1, 10, 32, 32, 80, 160.) The collapsed adjacency matrices are

$$\begin{pmatrix} 0 & 10 & 0 & 0 & 0 \\ 1 & 1 & 0 & 8 & 0 \\ 0 & 0 & 5 & 0 & 5 \\ 0 & 1 & 0 & 1 & 8 \\ 0 & 0 & 2 & 4 & 4 \end{pmatrix}$$

$$\begin{pmatrix} 0 & 0 & 64 & 0 & 0 \\ 0 & 0 & 32 & 0 & 32 \\ 1 & 5 & 8 & 20 & 30 \\ 0 & 0 & 16 & 16 & 32 \\ 0 & 2 & 12 & 16 & 34 \end{pmatrix}$$

$$\begin{pmatrix} 0 & 0 & 0 & 80 & 0 \\ 0 & 8 & 0 & 8 & 64 \\ 0 & 0 & 20 & 20 & 40 \\ 1 & 1 & 16 & 30 & 32 \\ 0 & 4 & 16 & 16 & 44 \end{pmatrix}$$

$$\begin{pmatrix} 0 & 0 & 0 & 0 & 160 \\ 0 & 0 & 32 & 64 & 64 \\ 0 & 5 & 30 & 40 & 85 \\ 0 & 8 & 32 & 32 & 88 \\ 1 & 4 & 34 & 44 & 77 \end{pmatrix}.$$

The only non-complete distance-regular generalized orbital graph is

$\Gamma\{2\}$, $\iota = \{10, 8, 8, 2; 1, 1, 4, 5\}$, G acts distance-transitively.

We may take the vertex-set of $\Gamma\{2\}$ to be the conjugacy class $2A$ of J_2.

Then vertices x, y are joined by an edge in $\Gamma\{2\}$ precisely when xy has order 2. This graph is described in more detail in [BCN89].

4.10 The Mathieu group M_{23}

A presentation for M_{23} is

$$\langle a3b5c3d3f6e, a4e3c4f \mid a = (cf)^2, b = (ef)^3,$$

$$1 = (eab)^3 = (bce)^5[= (bcd)^5] = (aecd)^4 = (bcef)^4 \rangle$$

(see [Soi85, Soi88]), in which

$$M_{22} \cong \langle a, b, c, d, e \rangle,$$

$$L_3(4) : 2_2 \cong \langle a, b, c, f, a^{bcde} \rangle,$$

$$2^4 : A_7 \cong \langle a, b, c, e, f \rangle,$$

$$A_8 \cong \langle a, b, d, e, f, e^{fdc} \rangle,$$

and

$$M_{11} \cong \langle a, b, c, d, f \rangle.$$

4.10.1 rank 2

(i) $\pi = \chi_1 + \chi_2$ of degree $23 = 1 + 22$.

This character corresponds to the unique 2-transitive representation of G of degree 23. A point stabilizer is $H \cong M_{22}$.

In the remainder of this section, let X be the set of points of size 23 in this representation. Then G preserves a Steiner system $S(4, 7, 23)$ with point-set X and block-set $\mathcal{B}$ consisting of 253 subsets of X of size 7, called *heptads*.

4.10.2 rank 3

(ii) $\pi = \chi_1 + \chi_2 + \chi_5$ of degree $253 = 1 + 22 + 230$.

There are two conjugacy classes of maximal subgroups of G of index

253 with associated permutation character π. The point stabilizers are $H \cong L_3(4)\colon 2_2$ and $H \cong 2^4\colon A_7$ in the two representations.

In the former case, H is the stabilizer of an unordered pair of points of X, and the subdegrees are 1, 42, 210. The collapsed adjacency matrices are

$$\begin{pmatrix} 0 & 42 & 0 \\ 1 & 21 & 20 \\ 0 & 4 & 38 \end{pmatrix}$$

$$\begin{pmatrix} 0 & 0 & 210 \\ 0 & 20 & 190 \\ 1 & 38 & 171 \end{pmatrix}.$$

In the latter case, where $H \cong 2^4\colon A_7$, H is the stabilizer of a heptad, that is, the action of G on Ω is permutationally isomorphic to its action on $\mathcal{B}$. The subdegrees are 1, 112, 140, and the collapsed adjacency matrices are

$$\begin{pmatrix} 0 & 112 & 0 \\ 1 & 36 & 75 \\ 0 & 60 & 52 \end{pmatrix}$$

$$\begin{pmatrix} 0 & 0 & 140 \\ 0 & 75 & 65 \\ 1 & 52 & 87 \end{pmatrix}.$$

If there were also an imprimitive representation of G with permutation character π, then we would have $H < L < G$ with $|G : L|$ dividing 253. Then $m = 23$ and H would have index 11 in $L \cong M_{22}$. There is no such subgroup H.

4.10.3 rank 4

(iii) $\pi = \chi_1 + \chi_2 + \chi_5 + \chi_9$ of degree $506 = 1 + 22 + 230 + 253$.

There is a unique class of maximal subgroups of G of index 506 with associated permutation character π. A point stabilizer is $H \cong A_8$.

The group M_{23} is the stabilizer of a point x in the 5-transitive representation of M_{24} of degree 24, and has two orbits on the set of 759 octads (the blocks of an $S(5,8,24)$) preserved by M_{24}, namely, the sets of octads which do, and which do not, contain the point x. The subgroup H is the stabilizer of an octad which does not contain x.

The subdegrees are 1, 15, 210, 280, and the collapsed adjacency matrices are

$$\begin{pmatrix} 0 & 15 & 0 & 0 \\ 1 & 0 & 14 & 0 \\ 0 & 1 & 2 & 12 \\ 0 & 0 & 9 & 6 \end{pmatrix}$$

$$\begin{pmatrix} 0 & 0 & 210 & 0 \\ 0 & 14 & 28 & 168 \\ 1 & 2 & 111 & 96 \\ 0 & 9 & 72 & 129 \end{pmatrix}$$

$$\begin{pmatrix} 0 & 0 & 0 & 280 \\ 0 & 0 & 168 & 112 \\ 0 & 12 & 96 & 172 \\ 1 & 6 & 129 & 144 \end{pmatrix}.$$

The only non-complete distance-regular generalized orbital graph is

$$\Gamma\{2\}, \quad \iota = \{15,14,12;1,1,9\}, \quad G \text{ acts distance-transitively.}$$

We may take the vertices of $\Gamma\{2\}$ to be the 506 octads of an $S(5,8,24)$ which do not contain a given point. We join two such octads by an edge in $\Gamma\{2\}$ precisely when they are disjoint. This graph is described in more detail in [BCN89].

If there were also an imprimitive permutation representation of G with character π, then we would have $H < L < G$ with $|G : L|$ a proper divisor of 506. Then either $m = 23$ and H has index 22 in $L \cong M_{22}$, or $m = 253$ and H has index 2 in $L \cong L_3(4)\colon 2$ or $2^4\colon A_7$. If $m = 23$, then H would be the stabilizer of an ordered pair of distinct points of X, and by Lemma 3.4 the representation would have rank at least 7, which is a contradiction. Hence $m = 253$. The subgroup $L \cong L_3(4)\colon 2$ has a unique subgroup of index 2, namely the stabilizer of an ordered pair of distinct points of X, and we have shown that H cannot be this subgroup. Hence $L \cong 2^4\colon A_7$, but this subgroup has no subgroup of index 2. Thus there are no imprimitive representations with associated character π.

(*iv*) $\pi = \chi_1 + \chi_2 + \chi_5 + \chi_{16}$ of degree $1288 = 1 + 22 + 230 + 1035$.

There is a unique class of maximal subgroups of G of index 1288 with associated permutation character π. A point stabilizer is $H \cong M_{11}$.

The subdegrees are 1, 165, 330, 792, and the collapsed adjacency matrices are

$$\begin{pmatrix} 0 & 165 & 0 & 0 \\ 1 & 0 & 92 & 72 \\ 0 & 46 & 11 & 108 \\ 0 & 15 & 45 & 105 \end{pmatrix}$$

$$\begin{pmatrix} 0 & 0 & 330 & 0 \\ 0 & 92 & 22 & 216 \\ 1 & 11 & 138 & 180 \\ 0 & 45 & 75 & 210 \end{pmatrix}$$

$$\begin{pmatrix} 0 & 0 & 0 & 792 \\ 0 & 72 & 216 & 504 \\ 0 & 108 & 180 & 504 \\ 1 & 105 & 210 & 476 \end{pmatrix}.$$

The non-complete distance-regular generalized orbital graphs are

$$\Gamma\{2,3\}, \quad \iota = \{495, 288; 1, 180\} \quad \text{(and complement)}.$$

The graph $\Gamma\{2,3\}$ has a distance-transitive (rank 3) action by M_{24}.

If there were also an imprimitive permutation representation of G with character π, then we would have $H < L < G$ with $|G : L|$ a proper divisor of 1288. Then $m = 23$, and H would have index 56 in $L \cong M_{22}$. However, M_{22} has no subgroup of this index.

4.10.4 rank 5

(*v*) $\pi = \chi_1 + \chi_2 + \chi_5 + \chi_9 + \chi_{17}$, of degree $2530 = 1 + 22 + 230 + 253 + 2024$.

Now G has no maximal subgroup of index 2530, and hence $H < L < G$ with $m := |G : L|$ dividing 2530, where m is one of 23, 253 or 506, and H has index 110, 10 or 5 in M_{22}, $L_3(4){:}2$ or A_8, respectively. There is no such subgroup H in any of these cases. Thus π is not a permutation character.

4.11 The Higman-Sims group HS

A presentation for HS is

$$\langle b4f3a3b5c3d, a4e3c \mid d = (bf)^2 = (eaf)^3, 1 = (eab)^3 = (bce)^5 [= (abc)^5] \rangle$$

(see [Soi85, Soi88]), in which

$$U_3(5):2 \cong \langle b, c, d, f, d^{cbae}, abfae \rangle > \langle b, c, d, f, d^{cbae} \rangle \cong U_3(5),$$

$$M_{22} \cong \langle a, b, c, d, e \rangle,$$

and

$$L_3(4):2_1 \cong \langle a, b, c, a^{bcde}, eafabceafbcbeaf \rangle.$$

A presentation for $HS:2$ is obtained from that of HS, by adjoining a generator t, and the relations

$$1 = t^2 = (at)^2 = (ct)^2 = (dt)^2 = (et)^2 = (ft)^2 = b^t be.$$

In this $HS:2$, we have

$$S_8 \times 2 \cong C(t) = \langle a, c, d, e, f, t, a^{ecb}, t^{(abcft)^8} \rangle.$$

Note also that t normalizes $\langle a, b, c, d, e \rangle \cong M_{22}$.

4.11.1 rank 2

(i) $\pi = \chi_1 + \chi_7$ of degree $176 = 1 + 175$.

The group G has two equivalent, but not permutationally isomorphic, 2-transitive representations with permutation character π, and these are interchanged by an outer automorphism of G. A point stabilizer is $H \cong U_3(5):2$.

4.11.2 rank 3

(ii) $\pi = \chi_1 + \chi_2 + \chi_3$ of degree $100 = 1 + 22 + 77$.

There is a unique class of subgroups of G of index 100. A representative is a maximal $H \cong M_{22}$, and the associated permutation character is π.

The subdegrees are 1, 22, 77, and the collapsed adjacency matrices are

$$\begin{pmatrix} 0 & 22 & 0 \\ 1 & 0 & 21 \\ 0 & 6 & 16 \end{pmatrix}$$

$$\begin{pmatrix} 0 & 0 & 77 \\ 0 & 21 & 56 \\ 1 & 16 & 60 \end{pmatrix}.$$

The orbital graph corresponding to the suborbit of length 22 is the Higman-Sims graph $\Gamma(HS)$, which was used by Higman and Sims [HS68] to construct their group. The automorphism group of $\Gamma(HS)$ is $HS:2$.

4.11.3 rank 4

(*iii*) $\pi = \chi_1 + \chi_2 + \chi_i + \chi_7$, where $i = 5$ or 6, of degree $352 = 1 + 22 + 154 + 175$.

The group G has no maximal subgroup of index 352, and hence $H < L < G$, with $m := |G : L|$ dividing 352. It follows that $m = 176$ and L belongs to one of the two conjugacy classes of subgroups in case (*i*) above. So $L \cong U_3(5):2$, and as L has a unique subgroup $H \cong U_3(5)$ of index 2, and H is not contained in a subgroup from the other class of subgroups of G of index 176, it follows that there are exactly two equivalent, but not permutationally isomorphic, representations of G of this type, corresponding to the two possibilities for π above (these two possibilities are interchanged by an outer automorphism of G (see ATLAS)). Since the representations of degree 176 are 2-transitive, it follows that each of the representations here has rank 4 and subdegrees 1, 1, 175, 175.

The collapsed adjacency matrices are

$$\begin{pmatrix} 0 & 1 & 0 & 0 \\ 1 & 0 & 0 & 0 \\ 0 & 0 & 0 & 1 \\ 0 & 0 & 1 & 0 \end{pmatrix}$$

$$\begin{pmatrix} 0 & 0 & 175 & 0 \\ 0 & 0 & 0 & 175 \\ 1 & 0 & 102 & 72 \\ 0 & 1 & 72 & 102 \end{pmatrix}$$

$$\begin{pmatrix} 0 & 0 & 0 & 175 \\ 0 & 0 & 175 & 0 \\ 0 & 1 & 72 & 102 \\ 1 & 0 & 102 & 72 \end{pmatrix}.$$

The non-complete distance-regular generalized orbital graphs are

$\Gamma\{3\}$, $\iota = \{175, 72, 1; 1, 72, 175\}$, G acts distance-transitively
$\Gamma\{4\}$, $\iota = \{175, 102, 1; 1, 102, 175\}$, G acts distance-transitively
$\Gamma\{3, 4\} \cong K_{176 \times 2}$.

$\Gamma\{3\}$ and $\Gamma\{4\}$ are well-known Taylor graphs, and are described in [BCN89, p. 370].

4.11.4 rank 5

(iv) $\pi = \chi_1 + \chi_2 + \chi_3 + \chi_7 + \chi_{13}$ of degree $1100 = 1 + 22 + 77 + 175 + 825$.

There is a unique class of maximal subgroups of G of index 1100 with associated permutation character π. A point stabilizer is $H \cong L_3(4){:}2_1$, the stabilizer in G of an unordered edge of the Higman-Sims graph $\Gamma(HS)$.

The subdegrees are 1, 42, 105, 280, 672, and the collapsed adjacency matrices are

$$\begin{pmatrix} 0 & 42 & 0 & 0 & 0 \\ 1 & 20 & 5 & 0 & 16 \\ 0 & 2 & 8 & 0 & 32 \\ 0 & 0 & 0 & 18 & 24 \\ 0 & 1 & 5 & 10 & 26 \end{pmatrix}$$

$$\begin{pmatrix} 0 & 0 & 105 & 0 & 0 \\ 0 & 5 & 20 & 0 & 80 \\ 1 & 8 & 0 & 64 & 32 \\ 0 & 0 & 24 & 9 & 72 \\ 0 & 5 & 5 & 30 & 65 \end{pmatrix}$$

$$\begin{pmatrix} 0 & 0 & 0 & 280 & 0 \\ 0 & 0 & 0 & 120 & 160 \\ 0 & 0 & 64 & 24 & 192 \\ 1 & 18 & 9 & 108 & 144 \\ 0 & 10 & 30 & 60 & 180 \end{pmatrix}$$

$$\begin{pmatrix} 0 & 0 & 0 & 0 & 672 \\ 0 & 16 & 80 & 160 & 416 \\ 0 & 32 & 32 & 192 & 416 \\ 0 & 24 & 72 & 144 & 432 \\ 1 & 26 & 65 & 180 & 400 \end{pmatrix}.$$

There are no non-complete distance-regular generalized orbital graphs.

If there were also an imprimitive representation of G with permutation character π, then we would have $H < L < G$ with $m := |G : L|$ dividing 1100. Then $m = 100$ and H would have index 11 in $L \cong M_{22}$. However there is no such subgroup.

(v) $\pi = \chi_1 + \chi_3 + \chi_4 + \chi_7 + \chi_9$ of degree $1100 = 1 + 77 + 154 + 175 + 693$.

There is a unique class of maximal subgroups of G of index 1100 with associated permutation character π. A point stabilizer is $H \cong S_8$.

The subdegrees are 1, 28, 105, 336, 630, and the collapsed adjacency matrices are

$$\begin{pmatrix} 0 & 28 & 0 & 0 & 0 \\ 1 & 0 & 15 & 12 & 0 \\ 0 & 4 & 0 & 0 & 24 \\ 0 & 1 & 0 & 12 & 15 \\ 0 & 0 & 4 & 8 & 16 \end{pmatrix}$$

$$\begin{pmatrix} 0 & 0 & 105 & 0 & 0 \\ 0 & 15 & 0 & 0 & 90 \\ 1 & 0 & 24 & 32 & 48 \\ 0 & 0 & 10 & 35 & 60 \\ 0 & 4 & 8 & 32 & 61 \end{pmatrix}$$

$$\begin{pmatrix} 0 & 0 & 0 & 336 & 0 \\ 0 & 12 & 0 & 144 & 180 \\ 0 & 0 & 32 & 112 & 192 \\ 1 & 12 & 35 & 108 & 180 \\ 0 & 8 & 32 & 96 & 200 \end{pmatrix}$$

$$\begin{pmatrix} 0 & 0 & 0 & 0 & 630 \\ 0 & 0 & 90 & 180 & 360 \\ 0 & 24 & 48 & 192 & 366 \\ 0 & 15 & 60 & 180 & 375 \\ 1 & 16 & 61 & 200 & 352 \end{pmatrix}.$$

There are no non-complete distance-regular generalized orbital graphs.

Arguing as in the previous case, we see that there is no imprimitive permutation representation with character π.

4.12 Automorphism group $HS:2$ of HS

A presentation for $HS:2$ is given in the section for HS.

4.12.1 rank 2

G has no faithful pseudo-permutation character of rank 2.

4.12.2 rank 3

(i) $\pi = \chi_1 + \chi_3 + \chi_5$ of degree $100 = 1 + 22 + 77$.

There is a unique class of subgroups of G of index 100. A representative is a maximal $H \cong M_{22}:2$, and the associated permutation character is π. The restriction of this representation to HS is the representation in case (ii) for HS, and so the collapsed adjacency matrices are the same as for that representation.

4.12.3 rank 4

(ii) $\pi = \chi_1 + \chi_2 + \chi_{10} + \chi_{11}$ of degree $352 = 1 + 1 + 175 + 175$.

The only subgroups H of G of index 352 are such that $H \cong U_3(5):2$ and $H < HS < G$. There is only one G-conjugacy class of such subgroups, and the associated permutation character is π.

The subdegrees are 1, 50, 126, 175, and the collapsed adjacency matrices are

$$\begin{pmatrix} 0 & 50 & 0 & 0 \\ 1 & 0 & 0 & 49 \\ 0 & 0 & 0 & 50 \\ 0 & 14 & 36 & 0 \end{pmatrix}$$

$$\begin{pmatrix} 0 & 0 & 126 & 0 \\ 0 & 0 & 0 & 126 \\ 1 & 0 & 0 & 125 \\ 0 & 36 & 90 & 0 \end{pmatrix}$$

$$\begin{pmatrix} 0 & 0 & 0 & 175 \\ 0 & 49 & 126 & 0 \\ 0 & 50 & 125 & 0 \\ 1 & 0 & 0 & 174 \end{pmatrix}.$$

The non-complete distance-regular generalized orbital graphs are

$\Gamma\{2\}$, $\iota = \{50, 49, 36; 1, 14, 50\}$, $\quad G$ acts distance-transitively

$\Gamma\{3\}$, $\iota = \{126, 125, 36; 1, 90, 126\}$, $\quad G$ acts distance-transitively

$\Gamma\{2, 3\} \cong K_{2 \times 176}$.

The graph $\Gamma\{2\}$ is the incidence graph of the famous $2 - (176, 50, 14)$ design discovered by G. Higman [Hig69], and $\Gamma\{3\}$ is the incidence graph of the complement of this design.

4.12.4 rank 5

(iii) $\quad \pi = \chi_1 + \chi_3 + \chi_5 + \chi_{10} + \chi_{19}$ of degree $1100 = 1 + 22 + 77 + 175 + 825$.

There is a unique class of maximal subgroups with associated permutation character π, namely $H \cong L_3(4): 2^2$. The restriction of this representation to HS is the rank 5 representation in case (iv) for HS. The collapsed adjacency matrices are the same as for the representation in case (iv) for HS, and there are no associated non-complete distance-regular graphs.

If there were also an imprimitive representation with permutation character π, then we would have $H < L < G$ with $m := |G : L|$ dividing 1100. Then $m = 100$, and H would have index 11 in $L \cong M_{22}: 2$. No such subgroup exists.

(iv) $\pi = \chi_1 + \chi_5 + \chi_7 + \chi_{10} + \chi_{14}$ of degree $1100 = 1 + 77 + 154 + 175 + 693$.

There is a unique class of maximal subgroups with associated permutation character π, namely $H \cong S_8 \times 2$. The restriction of this representation to HS is the rank 5 representation in case (v) for HS. The collapsed adjacency matrices are the same as for the representation in case (v) for HS, and there are no associated non-complete distance-regular graphs.

Arguing as in the previous case, we see that there is no imprimitive representation with permutation character π.

4.13 The Janko group J_3

G has no faithful pseudo-permutation character of rank less than or equal to 5.

4.14 Automorphism group J_3: 2 of J_3

4.14.1 rank 2

G has no faithful pseudo-permutation character of rank 2.

4.14.2 rank 3

(i) $\pi = \chi_1 + \chi_2 + \chi_4$ of degree $648 = 1 + 1 + 646$.

G has no subgroup of index 648.

4.14.3 rank 4

G has no pseudo-permutation character of rank 4.

4.14.4 rank 5

(ii) $\pi = \chi_1 + \chi_2 + \chi_4 + \chi_5 + \chi_6$ of degree $1296 = 1 + 1 + 646 + 324 + 324$.

G has no subgroup of index 1296.

4.15 The Mathieu group M_{24}

A presentation for M_{24} is

$$\langle a3b5c3d3f6e, a4e3c4f4g4b \mid a = (cf)^2, e = (bg)^2, b = (ef)^3,$$

$$1 = (aecd)^4 = (baefg)^3 = (bcef)^4 [= (bcd)^5] \rangle$$

(see [Soi85, Soi88]), in which

$$M_{23} \cong \langle a, b, c, d, e, f \rangle,$$

$$M_{22} : 2 \cong \langle a, b, c, d, e, g \rangle,$$

$$2^4 : A_8 \cong \langle a, b, c, e, f, g \rangle,$$

and

$$L_3(4) : S_3 \cong \langle a, b, c, a^{bcde}, f, eabg \rangle.$$

Another presentation for M_{24} is

$$\langle a3b3c10d3e, b4f, d3g \mid a = (cd)^5 = (cde)^5,$$

$$f = (cdg)^5, e = (bcf)^3, 1 = (abf)^3 \rangle$$

(see [Soi85]), in which

$$M_{12} : 2 \cong \langle a, b, c, d, e, f \rangle > \langle ab, bc, d, e, f \rangle \cong M_{12},$$

$$2^6 : 3 \cdot S_6 \cong \langle a, c, d, e, f, g, bcdegdcb \rangle > \langle a, c, d, e, f, g, d^{bcdegdcb} \rangle \cong 2^6 : 3 \cdot A_6,$$

and

$$2^6 : (L_3(2) \times S_3) \cong \langle a, b, d, e, f, g, (edg)^{dcbdcdc} \rangle.$$

4.15.1 rank 2

(i) $\pi = \chi_1 + \chi_2$ of degree $24 = 1 + 23$.

This character corresponds to the unique 2-transitive representation of G of degree 24. A point stabilizer is $H \cong M_{23}$.

In the remainder of this section, let X be the set of points, of size 24, in this representation. Then G preserves a Steiner system $S(5, 8, 24)$ with point-set X and block-set $\mathcal{B}$ consisting of 759 subsets of X of size 8, called *octads*. The octads are the supports of the weight 8 codewords of the binary Golay code $\mathcal{C}$ of length 24 (see [CS88, Chapter 11]) whose automorphism group is G. The support of a weight 12 codeword in $\mathcal{C}$ is called a *dodecad*, and dodecads come in complementary pairs.

(*ii*) $\pi = \chi_1 + \chi_7$ of degree $253 = 1 + 252$.

There is no subgroup of G of index 253.

4.15.2   rank 3

(*iii*) $\pi = \chi_1 + \chi_2 + \chi_7$ of degree $276 = 1 + 23 + 252$.

There is a unique class of subgroups of G of index 276. A representative is a maximal $H \cong M_{22}\!:\!2$, the stabilizer of an unordered pair of points of X, and the subdegrees are 1, 44, 231. The associated permutation character is π.

The collapsed adjacency matrices are

$$\begin{pmatrix} 0 & 44 & 0 \\ 1 & 22 & 21 \\ 0 & 4 & 40 \end{pmatrix}$$

$$\begin{pmatrix} 0 & 0 & 231 \\ 0 & 21 & 210 \\ 1 & 40 & 190 \end{pmatrix}.$$

(*iv*) $\pi = \chi_1 + \chi_7 + \chi_{14}$ of degree $1288 = 1 + 252 + 1035$.

There is a unique class of subgroups of G of index 1288. A representative is a maximal $H \cong M_{12}\!:\!2$, the stabilizer of a complementary pair of dodecads. The associated permutation character is π.

The subdegrees are 1, 495, 792, and the collapsed adjacency matrices are

$$\begin{pmatrix} 0 & 495 & 0 \\ 1 & 206 & 288 \\ 0 & 180 & 315 \end{pmatrix}$$

$$\begin{pmatrix} 0 & 0 & 792 \\ 0 & 288 & 504 \\ 1 & 315 & 476 \end{pmatrix}.$$

(*v*) $\pi = \chi_1 + \chi_7 + \chi_{18}$ of degree $2024 = 1 + 252 + 1771$.

There is a unique class of maximal subgroups of G of index 2024, namely $H \cong L_3(4):S_3$. However the permutation character for the permutation representation associated with this subgroup H has rank 5. Thus we must have $H < L < G$ with $|G:L|$ dividing 2024. However G has no proper subgroup of index a proper divisor of 2024, so π is not a permutation character.

4.15.3 rank 4

(vi) $\quad \pi = \chi_1 + \chi_2 + \chi_7 + \chi_9$ of degree $759 = 1 + 23 + 252 + 483$.

There is a unique class of subgroups of G of index 759. A representative is a maximal $H \cong 2^4:A_8$, the stabilizer of an octad in $\mathcal{B}$. Thus the action of G on Ω is permutationally isomorphic to its action on $\mathcal{B}$. The associated permutation character is π.

The subdegrees are 1, 30, 280, 448, and the collapsed adjacency matrices are

$$\begin{pmatrix} 0 & 30 & 0 & 0 \\ 1 & 1 & 28 & 0 \\ 0 & 3 & 3 & 24 \\ 0 & 0 & 15 & 15 \end{pmatrix}$$

$$\begin{pmatrix} 0 & 0 & 280 & 0 \\ 0 & 28 & 28 & 224 \\ 1 & 3 & 140 & 136 \\ 0 & 15 & 85 & 180 \end{pmatrix}$$

$$\begin{pmatrix} 0 & 0 & 0 & 448 \\ 0 & 0 & 224 & 224 \\ 0 & 24 & 136 & 288 \\ 1 & 15 & 180 & 252 \end{pmatrix}.$$

The only non-complete distance-regular generalized orbital graph is

$$\Gamma\{2\}, \quad \iota = \{30, 28, 24; 1, 3, 15\}, \quad G \text{ acts distance-transitively.}$$

We may take the vertex-set of $\Gamma\{2\}$ to be the set $\mathcal{B}$ of 759 octads. We join two such octads by an edge in $\Gamma\{2\}$ precisely when they are disjoint. This graph is described in more detail in [BCN89, p. 366] where it is called the *large Witt graph*.

(*vii*) $\pi = \chi_1 + \chi_7 + \chi_9 + \chi_{14}$ of degree $1771 = 1 + 252 + 483 + 1035$.

There is a unique class of subgroups of G of index 1771. A representative is a maximal $H \cong 2^6\!:\!3\!\cdot\!S_6$. The associated permutation character is π.

The subdegrees are 1, 90, 240, 1440, and the collapsed adjacency matrices are

$$
\begin{pmatrix}
0 & 90 & 0 & 0 \\
1 & 17 & 24 & 48 \\
0 & 9 & 9 & 72 \\
0 & 3 & 12 & 75
\end{pmatrix}
$$

$$
\begin{pmatrix}
0 & 0 & 240 & 0 \\
0 & 24 & 24 & 192 \\
1 & 9 & 38 & 192 \\
0 & 12 & 32 & 196
\end{pmatrix}
$$

$$
\begin{pmatrix}
0 & 0 & 0 & 1440 \\
0 & 48 & 192 & 1200 \\
0 & 72 & 192 & 1176 \\
1 & 75 & 196 & 1168
\end{pmatrix}.
$$

There are no non-complete distance-regular generalized orbital graphs.

(*viii*) $\pi = \chi_1 + \chi_7 + \chi_{18} + \chi_{19}$ of degree $4048 = 1 + 252 + 1771 + 2024$.

There is no maximal subgroup of G of index 4048, and hence $H < L < G$ with $m := |G : L|$ a proper divisor of 4048. It follows that $m = 2024$ and H has index 2 in $L \cong L_3(4)\!:\!S_3$; however the action of G on the set of right cosets of L in G has rank 5, and so the action of G on Ω has rank at least 6, which is a contradiction.

(*ix*) $\pi = \chi_1 + \chi_7 + \chi_{14} + \chi_{25}$ of degree $7084 = 1 + 252 + 1035 + 5796$.

There is no maximal subgroup of G of index 7084, and hence $H < L < G$ with $m := |G : L|$ a proper divisor of 7084. It follows that $m = 1771$ and H has index 4 in $L \cong 2^6\!:\!3\!\cdot\!S_6$. However, the action of G on the set of right cosets of L in G is permutationally isomorphic to its rank 4 action in case (*vii*) above, and it follows that the action on Ω has rank greater than 4, which is a contradiction.

4.15.4 rank 5

(x) $\pi = \chi_1 + \chi_2 + \chi_7 + \chi_9 + \chi_{17}$, of degree $2024 = 1 + 23 + 252 + 483 + 1265$.

There is a unique class of subgroups of G of index 2024. A representative is a maximal $H \cong L_3(4){:}S_3$, the stabilizer of an unordered triple of points of X. The associated permutation character is π.

The subdegrees are 1, 63, 210, 630, 1120, and the collapsed adjacency matrices are

$$
\begin{pmatrix}
0 & 63 & 0 & 0 & 0 \\
1 & 22 & 0 & 40 & 0 \\
0 & 0 & 6 & 9 & 48 \\
0 & 4 & 3 & 40 & 16 \\
0 & 0 & 9 & 9 & 45
\end{pmatrix}
$$

$$
\begin{pmatrix}
0 & 0 & 210 & 0 & 0 \\
0 & 0 & 20 & 30 & 160 \\
1 & 6 & 32 & 75 & 96 \\
0 & 3 & 25 & 54 & 128 \\
0 & 9 & 18 & 72 & 111
\end{pmatrix}
$$

$$
\begin{pmatrix}
0 & 0 & 0 & 630 & 0 \\
0 & 40 & 30 & 400 & 160 \\
0 & 9 & 75 & 162 & 384 \\
1 & 40 & 54 & 247 & 288 \\
0 & 9 & 72 & 162 & 387
\end{pmatrix}
$$

$$
\begin{pmatrix}
0 & 0 & 0 & 0 & 1120 \\
0 & 0 & 160 & 160 & 800 \\
0 & 48 & 96 & 384 & 592 \\
0 & 16 & 128 & 288 & 688 \\
1 & 45 & 111 & 387 & 576
\end{pmatrix}.
$$

The only non-complete distance-regular generalized orbital graph is

$$\Gamma\{2\} \cong J(24,3), \quad \iota = \{63, 40, 19; 1, 4, 9\}.$$

(xi) $\pi = \chi_1 + \chi_2 + \chi_7 + \chi_{14} + \chi_{17}$ of degree $2576 = 1 + 23 + 252 + 1035 + 1265$.

There is no maximal subgroup of G of index 2576, and hence $H < L < G$ with $m := |G : L|$ a proper divisor of 2576. It follows that H has index

2 in $L \cong M_{12}:2$, the stabilizer of a complementary pair of dodecads in the representation in (iv) above. Therefore $H \cong M_{12}$, and the action of G on the cosets of H is permutationally isomorphic to the action of G on the 2576 dodecads contained in X.

Two dodecads meet in 12, 0, 4, 8, or 6 points, giving rise to suborbits of H of respective lengths 1, 1, 495, 495, 1584. The collapsed adjacency matrices are

$$\begin{pmatrix} 0 & 1 & 0 & 0 & 0 \\ 1 & 0 & 0 & 0 & 0 \\ 0 & 0 & 0 & 1 & 0 \\ 0 & 0 & 1 & 0 & 0 \\ 0 & 0 & 0 & 0 & 1 \end{pmatrix}$$

$$\begin{pmatrix} 0 & 0 & 495 & 0 & 0 \\ 0 & 0 & 0 & 495 & 0 \\ 1 & 0 & 22 & 184 & 288 \\ 0 & 1 & 184 & 22 & 288 \\ 0 & 0 & 90 & 90 & 315 \end{pmatrix}$$

$$\begin{pmatrix} 0 & 0 & 0 & 495 & 0 \\ 0 & 0 & 495 & 0 & 0 \\ 0 & 1 & 184 & 22 & 288 \\ 1 & 0 & 22 & 184 & 288 \\ 0 & 0 & 90 & 90 & 315 \end{pmatrix}$$

$$\begin{pmatrix} 0 & 0 & 0 & 0 & 1584 \\ 0 & 0 & 0 & 0 & 1584 \\ 0 & 0 & 288 & 288 & 1008 \\ 0 & 0 & 288 & 288 & 1008 \\ 1 & 1 & 315 & 315 & 952 \end{pmatrix}.$$

The only non-complete distance-regular generalized orbital graph is

$$\Gamma\{3,4,5\} \cong K_{1288\times2}.$$

(xii) $\pi = \chi_1 + \chi_7 + \chi_9 + \chi_{14} + \chi_{18}$ of degree $3542 = 1 + 252 + 483 + 1035 + 1771$.

There is no maximal subgroup of G of index 3542, and hence $H < L < G$ with $m := |G : L|$ a proper divisor of 3542. It follows that $m = 1771$, and H has index 2 in $L \cong 2^6 : 3 \cdot S_6$. Thus $H \cong 2^6 : 3 \cdot A_6$.

Direct calculation shows that the action of G on the cosets of H has rank 5, with subdegrees 1, 1, 180, 480, 2880, and the associated permutation character must be π, since π is the unique rank 5 pseudo-permutation character of G of degree 3542. The collapsed adjacency matrices are

$$\begin{pmatrix} 0 & 1 & 0 & 0 & 0 \\ 1 & 0 & 0 & 0 & 0 \\ 0 & 0 & 1 & 0 & 0 \\ 0 & 0 & 0 & 1 & 0 \\ 0 & 0 & 0 & 0 & 1 \end{pmatrix}$$

$$\begin{pmatrix} 0 & 0 & 180 & 0 & 0 \\ 0 & 0 & 180 & 0 & 0 \\ 1 & 1 & 34 & 48 & 96 \\ 0 & 0 & 18 & 18 & 144 \\ 0 & 0 & 6 & 24 & 150 \end{pmatrix}$$

$$\begin{pmatrix} 0 & 0 & 0 & 480 & 0 \\ 0 & 0 & 0 & 480 & 0 \\ 0 & 0 & 48 & 48 & 384 \\ 1 & 1 & 18 & 76 & 384 \\ 0 & 0 & 24 & 64 & 392 \end{pmatrix}$$

$$\begin{pmatrix} 0 & 0 & 0 & 0 & 2880 \\ 0 & 0 & 0 & 0 & 2880 \\ 0 & 0 & 96 & 384 & 2400 \\ 0 & 0 & 144 & 384 & 2352 \\ 1 & 1 & 150 & 392 & 2336 \end{pmatrix}.$$

The only non-complete distance-regular generalized orbital graph is

$$\Gamma\{3,4,5\} \cong K_{1771\times2}.$$

(*xiii*) $\pi = \chi_1 + \chi_7 + \chi_9 + \chi_{14} + \chi_{19}$ of degree $3795 = 1 + 252 + 483 + 1035 + 2024$.

There is a unique class of maximal subgroups of G of index 3795. A representative is $H \cong 2^6 : (L_3(2) \times S_3)$. The associated permutation character is π.

The subdegrees are 1, 42, 56, 1008, 2688, and the collapsed adjacency matrices are

$$\begin{pmatrix} 0 & 42 & 0 & 0 & 0 \\ 1 & 13 & 4 & 24 & 0 \\ 0 & 3 & 3 & 36 & 0 \\ 0 & 1 & 2 & 15 & 24 \\ 0 & 0 & 0 & 9 & 33 \end{pmatrix}$$

$$\begin{pmatrix} 0 & 0 & 56 & 0 & 0 \\ 0 & 4 & 4 & 48 & 0 \\ 1 & 3 & 4 & 0 & 48 \\ 0 & 2 & 0 & 14 & 40 \\ 0 & 0 & 1 & 15 & 40 \end{pmatrix}$$

$$\begin{pmatrix} 0 & 0 & 0 & 1008 & 0 \\ 0 & 24 & 48 & 360 & 576 \\ 0 & 36 & 0 & 252 & 720 \\ 1 & 15 & 14 & 258 & 720 \\ 0 & 9 & 15 & 270 & 714 \end{pmatrix}$$

$$\begin{pmatrix} 0 & 0 & 0 & 0 & 2688 \\ 0 & 0 & 0 & 576 & 2112 \\ 0 & 0 & 48 & 720 & 1920 \\ 0 & 24 & 40 & 720 & 1904 \\ 1 & 33 & 40 & 714 & 1900 \end{pmatrix}.$$

There are no non-complete distance-regular generalized orbital graphs.

If there were also an imprimitive representation of G with permutation character π, then we would have $H < L < G$ with $m := |G : L|$ a proper divisor of 3795. Then $m = 759$ and H has index 5 in $L \cong 2^4 : A_8$. However L has no subgroup of index 5.

(*xiv*) $\pi = \chi_1 + \chi_7 + \chi_{14} + \chi_{19} + \chi_{25}$ of degree $9108 = 1 + 252 + 1035 + 2024 + 5796$.

There is no maximal subgroup of G of index 9108, and hence $H < L < G$ with $m := |G : L|$ a proper divisor of 9108. Then m is either 276 or 759 and H would have index 33 or 12 in $L \cong M_{22} : 2$ or $2^4 : A_8$ respectively. However neither of these subgroups L has a subgroup of the required index. Thus π is not a permutation character.

4.16 The McLaughlin group McL

A presentation for McL is

$$\langle a3b5c3d3f6e, a4e3c4f \mid a = (cf)^2, b = (ef)^3,$$

$$1 = (eab)^3 = (bce)^5 [= (bcd)^5] = (aecd)^4 = (cef)^7 \rangle$$

(see [Soi85, Soi88]), in which

$$U_4(3) \cong \langle cef, dfcd \rangle,$$

$$M_{22} \cong \langle a, b, c, d, e \rangle,$$

$$U_3(5) \cong \langle b, (de)^{cbac}, acdcbeace, (cdfbeaecd)^3 \rangle,$$

and

$$3^{1+4}_+ : 2.S_5 \cong N(\langle (cdfbeaecd)^{10} \rangle) = \langle cdfbeaecd, cbadecbcbadcecbaecd \rangle.$$

The short words generating $U_4(3)$ were found by M. Schönert.

4.16.1 rank 2

G has no pseudo-permutation character of rank 2.

4.16.2 rank 3

(i) $\pi = \chi_1 + \chi_2 + \chi_4$ of degree $275 = 1 + 22 + 252$.

There is a unique class of subgroups of G of index 275. A representative is a maximal $H \cong U_4(3)$, and the associated permutation character is π.

The subdegrees are 1, 112, 162, and the collapsed adjacency matrices are

$$\begin{pmatrix} 0 & 112 & 0 \\ 1 & 30 & 81 \\ 0 & 56 & 56 \end{pmatrix}$$

$$\begin{pmatrix} 0 & 0 & 162 \\ 0 & 81 & 81 \\ 1 & 56 & 105 \end{pmatrix}.$$

The orbital graph corresponding to the suborbit of length 112 is the

McLaughlin graph $\Gamma(McL)$, and McLaughlin [McL69] constructed his group $G = McL$ as a group of automorphisms of this graph. The automorphism group of $\Gamma(McL)$ is Aut $(McL) = McL{:}2$.

4.16.3 rank 4

(ii) $\pi = \chi_1 + \chi_2 + \chi_4 + \chi_9$ of degree $2025 = 1 + 22 + 252 + 1750$.

There are just two classes of subgroups of G of index 2025, and these are interchanged by an outer automorphism of G. For each class, a representative is a maximal $H \cong M_{22}$, and the associated permutation character is π.

For each of these two equivalent, but not permutationally isomorphic, representations, the subdegrees are 1, 330, 462, 1232, and the collapsed adjacency matrices are

$$\begin{pmatrix} 0 & 330 & 0 & 0 \\ 1 & 7 & 154 & 168 \\ 0 & 110 & 20 & 200 \\ 0 & 45 & 75 & 210 \end{pmatrix}$$

$$\begin{pmatrix} 0 & 0 & 462 & 0 \\ 0 & 154 & 28 & 280 \\ 1 & 20 & 185 & 256 \\ 0 & 75 & 96 & 291 \end{pmatrix}$$

$$\begin{pmatrix} 0 & 0 & 0 & 1232 \\ 0 & 168 & 280 & 784 \\ 0 & 200 & 256 & 776 \\ 1 & 210 & 291 & 730 \end{pmatrix}.$$

There are no non-complete distance-regular generalized orbital graphs.

(iii) $\pi = \chi_1 + \chi_2 + \chi_4 + \chi_{20}$ of degree $9900 = 1 + 22 + 252 + 9625$.

There are no maximal subgroups of G of index 9900, and hence $H < L < G$ where $m := |G : L|$ divides 9900. Then $m = 275$, and H has index 36 in $L \cong M_{22}$. However there is no such subgroup, and hence π is not a permutation character.

4.16.4 rank 5

(iv) $\pi = \chi_1 + \chi_2 + \chi_4 + \chi_9 + \chi_{14}$ of degree $7128 = 1 + 22 + 252 + 1750 + 5103$.

There is a unique class of subgroups of G of index 7128. A representative is a maximal $H \cong U_3(5)$, and the associated permutation character is π.

The subdegrees are 1, 252, 750, 2625, 3500, and the collapsed adjacency matrices are

$$
\begin{pmatrix}
0 & 252 & 0 & 0 & 0 \\
1 & 1 & 125 & 0 & 125 \\
0 & 42 & 0 & 168 & 42 \\
0 & 0 & 48 & 48 & 156 \\
0 & 9 & 9 & 117 & 117
\end{pmatrix}
$$

$$
\begin{pmatrix}
0 & 0 & 750 & 0 & 0 \\
0 & 125 & 0 & 500 & 125 \\
1 & 0 & 224 & 77 & 448 \\
0 & 48 & 22 & 376 & 304 \\
0 & 9 & 96 & 228 & 417
\end{pmatrix}
$$

$$
\begin{pmatrix}
0 & 0 & 0 & 2625 & 0 \\
0 & 0 & 500 & 500 & 1625 \\
0 & 168 & 77 & 1316 & 1064 \\
1 & 48 & 376 & 808 & 1392 \\
0 & 117 & 228 & 1044 & 1236
\end{pmatrix}
$$

$$
\begin{pmatrix}
0 & 0 & 0 & 0 & 3500 \\
0 & 125 & 125 & 1625 & 1625 \\
0 & 42 & 448 & 1064 & 1946 \\
0 & 156 & 304 & 1392 & 1648 \\
1 & 117 & 417 & 1236 & 1729
\end{pmatrix}.
$$

There are no non-complete distance-regular generalized orbital graphs.

(v) $\pi = \chi_1 + \chi_4 + \chi_{12} + \chi_{14} + \chi_{15}$ of degree $15400 = 1 + 252 + 4500 + 5103 + 5544$.

There are two conjugacy classes of maximal subgroups of G of index 15400, with representatives $3^4 : M_{10}$ and $N(3A) \cong 3^{1+4}_+ : 2.S_5$.

Suppose $H \cong 3^4 : M_{10}$. Then H is the stabilizer in McL of an unordered edge of the McLaughlin graph $\Gamma = \Gamma(McL)$ defined in (i) above (G is transitive on the edges of Γ). It is easy to show that there are more than five isomorphism types of subgraphs of Γ induced on the unions of (not necessarily distinct) pairs of unordered edges, and so the rank of G on the cosets of H is greater than 5. In fact, direct calculation shows this rank to be 10.

Now suppose $H \cong 3^{1+4}_+ : 2.S_5$. Then H is the normalizer of a $3A$ subgroup of order 3, and is also the stabilizer in McL of a 5-clique of $\Gamma(McL)$.

Direct calculation shows that G has rank 5, and so the associated permutation character must be π, the unique rank 5 pseudo-permutation character of G of degree 15400. The subdegrees are 1, 90, 1215, 2430, 11664, and the collapsed adjacency matrices are

$$\begin{pmatrix} 0 & 90 & 0 & 0 & 0 \\ 1 & 8 & 54 & 27 & 0 \\ 0 & 4 & 32 & 6 & 48 \\ 0 & 1 & 3 & 14 & 72 \\ 0 & 0 & 5 & 15 & 70 \end{pmatrix}$$

$$\begin{pmatrix} 0 & 0 & 1215 & 0 & 0 \\ 0 & 54 & 432 & 81 & 648 \\ 1 & 32 & 246 & 168 & 768 \\ 0 & 3 & 84 & 192 & 936 \\ 0 & 5 & 80 & 195 & 935 \end{pmatrix}$$

$$\begin{pmatrix} 0 & 0 & 0 & 2430 & 0 \\ 0 & 27 & 81 & 378 & 1944 \\ 0 & 6 & 168 & 384 & 1872 \\ 1 & 14 & 192 & 351 & 1872 \\ 0 & 15 & 195 & 390 & 1830 \end{pmatrix}$$

$$\begin{pmatrix} 0 & 0 & 0 & 0 & 11664 \\ 0 & 0 & 648 & 1944 & 9072 \\ 0 & 48 & 768 & 1872 & 8976 \\ 0 & 72 & 936 & 1872 & 8784 \\ 1 & 70 & 935 & 1830 & 8828 \end{pmatrix}.$$

There are no non-complete distance-regular generalized orbital graphs.

If there were an imprimitive representation of G with permutation character π, we would have $H < L < G$ with $m := |G : L|$ dividing 15400.

Then $m = 275$ and H would have index 56 in $L \cong U_4(3)$. However L has no such subgroup.

4.17 Automorphism group McL: 2 of McL

We do not give a presentation for McL: 2. All faithful permutation representations of rank at most 5 for McL: 2 are extensions of representations of the same rank for McL.

4.17.1 rank 2

G has no faithful pseudo-permutation character of rank 2.

4.17.2 rank 3

(i) $\pi = \chi_1 + \chi_3 + \chi_7$ of degree $275 = 1 + 22 + 252$.

There is a unique class of subgroups of G of index 275. A representative is a maximal $H \cong U_4(3): 2_3$, and the associated permutation character is π. The restriction of this representation to McL is the representation in case (i) for McL, and so the collapsed adjacency matrices are the same as for that representation.

4.17.3 rank 4

G has no pseudo-permutation character of rank 4.

4.17.4 rank 5

(ii) $\pi = \chi_1 + \chi_4 + \chi_7 + \chi_{14} + \chi_{24}$ of degree $7128 = 1 + 22 + 252 + 1750 + 5103$.

There is a unique class of subgroups of G of index 7128. A representative is a maximal $H \cong U_3(5): 2$, and the associated permutation character is π. The restriction of this representation to McL is the representation in case (iv) for McL, and so the collapsed adjacency matrices are the same as for that representation. There are no non-complete distance-regular generalized orbital graphs.

(iii) $\pi = \chi_1 + \chi_7 + \chi_{20} + \chi_{24} + \chi_{26}$ of degree $15400 = 1 + 252 + 4500 + 5103 + 5544$.

There are just two conjugacy classes of maximal subgroups of G of index 15400, with representatives $3^4:(M_{10} \times 2)$ and $N(3A) \cong 3_+^{1+4}:4.S_5$.

Arguing as in case (v) for McL, we see that there is exactly one conjugacy class of maximal subgroups of $McL:2$ of index 15400 with associated permutation character π. A representative is $H \cong 3_+^{1+4}:4.S_5$, and the restriction to McL is the rank 5 representation of case (v) of McL. Thus the collapsed adjacency matrices are the same as for that representation, and there are no associated non-complete distance-regular graphs.

If there were an imprimitive representation of G with permutation character π, we would have $H < L < G$ with $m := |G : L|$ dividing 15400. Then $m = 275$ and H has index 56 in $L \cong U_4(3):2_3$. However L has no such subgroup.

4.18 The Held group He

A presentation for He is

$$\langle a4b3c10d3e, d3f4g3d \mid e = (fg)^2 = (abc)^3, a = (cd)^5, 1 = (cdfg)^4\rangle$$

(see [Soi91], where also words for generators of many maximal subgroups are given). In this He, we have

$$S_4(4):2 \cong \langle a, c, d, e, f, (ab)^2, (bcde)^7\rangle.$$

A presentation for $He:2$ is obtained from this presentation for He by adjoining a generator t, and the relations

$$1 = t^2 = (at)^2 = (bt)^2 = (ct)^2 = (dt)^2 = (et)^2 = f^t g.$$

In this $He:2$, we have

$$S_4(4):4 \cong \langle a, c, d, e, f, (ab)^2, (bcde)^7, bcdcdgfdcbdft\rangle.$$

4.18.1 rank 2

G has no pseudo-permutation character of rank 2.

4.18.2 rank 3

G has no pseudo-permutation character of rank 3.

4.18.3 rank 4

G has no pseudo-permutation character of rank 4.

4.18.4 rank 5

(i) $\pi = \chi_1 + \chi_2 + \chi_3 + \chi_6 + \chi_9$ of degree $2058 = 1 + 51 + 51 + 680 + 1275$.

There is a unique class of subgroups of G of index 2058. A representative is a maximal $H \cong S_4(4){:}2$, and the associated permutation character is π.

The subdegrees are 1, 136, 136, 425, 1360, and the collapsed adjacency matrices are

$$\begin{pmatrix} 0 & 136 & 0 & 0 & 0 \\ 0 & 0 & 36 & 0 & 100 \\ 1 & 0 & 0 & 75 & 60 \\ 0 & 24 & 0 & 16 & 96 \\ 0 & 6 & 10 & 30 & 90 \end{pmatrix}$$

$$\begin{pmatrix} 0 & 0 & 136 & 0 & 0 \\ 1 & 0 & 0 & 75 & 60 \\ 0 & 36 & 0 & 0 & 100 \\ 0 & 0 & 24 & 16 & 96 \\ 0 & 10 & 6 & 30 & 90 \end{pmatrix}$$

$$\begin{pmatrix} 0 & 0 & 0 & 425 & 0 \\ 0 & 75 & 0 & 50 & 300 \\ 0 & 0 & 75 & 50 & 300 \\ 1 & 16 & 16 & 136 & 256 \\ 0 & 30 & 30 & 80 & 285 \end{pmatrix}$$

$$\begin{pmatrix} 0 & 0 & 0 & 0 & 1360 \\ 0 & 60 & 100 & 300 & 900 \\ 0 & 100 & 60 & 300 & 900 \\ 0 & 96 & 96 & 256 & 912 \\ 1 & 90 & 90 & 285 & 894 \end{pmatrix}.$$

There are no non-complete distance-regular generalized orbital graphs.

4.19 Automorphism group $He{:}2$ of He

A presentation for $He{:}2$ is given in the section for He.

4.19.1 rank 2

G has no faithful pseudo-permutation character of rank 2.

4.19.2 rank 3

G has no pseudo-permutation character of rank 3.

4.19.3 rank 4

(i) $\pi = \chi_1 + \chi_3 + \chi_5 + \chi_9$ of degree $2058 = 1 + 102 + 680 + 1275$.

There is a unique class of subgroups of G of index 2058. A representative is a maximal $H \cong S_4(4){:}4$. The restriction of this representation to He is the rank 5 representation of case (i) for He. Hence G has rank 4 or 5 on Ω. However, since (see the next subsection) G has no rank 5 permutation character, it follows that G has rank 4, permutation character π, and subdegrees 1, 272, 425, 1360.

The collapsed adjacency matrices are

$$\begin{pmatrix} 0 & 272 & 0 & 0 \\ 1 & 36 & 75 & 160 \\ 0 & 48 & 32 & 192 \\ 0 & 32 & 60 & 180 \end{pmatrix}$$

$$\begin{pmatrix} 0 & 0 & 425 & 0 \\ 0 & 75 & 50 & 300 \\ 1 & 32 & 136 & 256 \\ 0 & 60 & 80 & 285 \end{pmatrix}$$

$$\begin{pmatrix} 0 & 0 & 0 & 1360 \\ 0 & 160 & 300 & 900 \\ 0 & 192 & 256 & 912 \\ 1 & 180 & 285 & 894 \end{pmatrix}.$$

There are no non-complete distance-regular generalized orbital graphs.

4.19.4 rank 5

G has no pseudo-permutation character of rank 5.

4.20 The Rudvalis group Ru

In [Wei91], R. Weiss gives a presentation for $G = Ru$, and he also gives words generating a subgroup $^2F_4(2)'$ and words generating $^2F_4(2)$. His presentation is derived in order to classify a certain geometry for G. We remark that permutations of degree 4060 generating G are available in the MAGMA system [CP95].

4.20.1 rank 2

G has no pseudo-permutation character of rank 2.

4.20.2 rank 3

(i) $\pi = \chi_1 + \chi_5 + \chi_6$ of degree $4060 = 1 + 783 + 3276$.

There is a unique class of subgroups of G of index 4060. A representative is a maximal $H \cong {}^2F_4(2)$, and the associated permutation character is π.

The subdegrees are 1, 1755, 2304, and the collapsed adjacency matrices are

$$\begin{pmatrix} 0 & 1755 & 0 \\ 1 & 730 & 1024 \\ 0 & 780 & 975 \end{pmatrix}$$

$$\begin{pmatrix} 0 & 0 & 2304 \\ 0 & 1024 & 1280 \\ 1 & 975 & 1328 \end{pmatrix}.$$

4.20.3 rank 4

G has no pseudo-permutation character of rank 4.

4.20.4 rank 5

(ii) $\pi = \chi_1 + \chi_4 + \chi_5 + \chi_6 + \chi_7$ of degree $8120 = 1 + 406 + 783 + 3276 + 3654$.

There is no maximal subgroup of index 8120, and hence $H < L < G$ with $m := |G : L|$ dividing 8120. It follows that $m = 4060$ and H has index 2 in $L \cong {}^2F_4(2)$. Hence $H \cong {}^2F_4(2)'$. Thus G preserves a block system Σ consisting of 4060 blocks of size 2, and the action of G on Σ is permutationally isomorphic to the rank 3 representation in (i) above with subdegrees 1, 1755, 2304. Let B be the block of Σ containing α. Then H fixes both points of B, and (by the ATLAS, p.74) H is transitive on the two L-orbits, Σ_1 and Σ_2, in $\Sigma \setminus \{B\}$ of lengths 1755 and 2304 respectively. Let $B_i \in \Sigma_i$ for $i = 1, 2$. Then $H_{B_1} \cong 2.[2^8]:5:4$ and $H_{B_2} \cong L_2(25)$ (by the ATLAS, p.74). Since H_{B_2} is the unique subgroup of L_{B_2} of index 2, it follows that H_{B_2} fixes B_2 pointwise. Hence H has two orbits of length 2304 on the points of Ω in blocks of Σ_2. Now H must either be transitive on the 3510 points of Ω in blocks of Σ_1, or have two orbits of length 1755 on this set. Thus G has rank 5 or 6. Suppose that G has rank 6. Then G must have a permutation character π' of rank 6, degree 8120, which contains the character of case (i), such that the difference between π' and this character is the sum of three nontrivial irreducible characters. By the ATLAS, p.127, this difference would have to be $\chi_i + \chi_4 + \chi_6$, where $i = 2$ or $i = 3$. However, in this case χ_6 would have multiplicity 2 in π' which would force G to have rank 8, which is a contradiction. Hence G has rank 5, subdegrees 1, 1, 2304, 2304, 3510, and associated permutation character π.

The collapsed adjacency matrices are

$$\begin{pmatrix} 0 & 1 & 0 & 0 & 0 \\ 1 & 0 & 0 & 0 & 0 \\ 0 & 0 & 0 & 1 & 0 \\ 0 & 0 & 1 & 0 & 0 \\ 0 & 0 & 0 & 0 & 1 \end{pmatrix}$$

$$\begin{pmatrix} 0 & 0 & 2304 & 0 & 0 \\ 0 & 0 & 0 & 2304 & 0 \\ 1 & 0 & 728 & 600 & 975 \\ 0 & 1 & 600 & 728 & 975 \\ 0 & 0 & 640 & 640 & 1024 \end{pmatrix}$$

$$\begin{pmatrix} 0 & 0 & 0 & 2304 & 0 \\ 0 & 0 & 2304 & 0 & 0 \\ 0 & 1 & 600 & 728 & 975 \\ 1 & 0 & 728 & 600 & 975 \\ 0 & 0 & 640 & 640 & 1024 \end{pmatrix}$$

$$\begin{pmatrix} 0 & 0 & 0 & 0 & 3510 \\ 0 & 0 & 0 & 0 & 3510 \\ 0 & 0 & 975 & 975 & 1560 \\ 0 & 0 & 975 & 975 & 1560 \\ 1 & 1 & 1024 & 1024 & 1460 \end{pmatrix}.$$

The only non-complete distance-regular generalized orbital graph is

$$\Gamma\{3,4,5\} \cong K_{4060 \times 2}.$$

4.21 The Suzuki group *Suz*

A presentation for *Suz* is

$$\langle a5b3c8d3e, b3f4g3b \mid a = (cd)^4 = (fg)^2,$$

$$1 = (abcf)^5[= (abcg)^5] = (bfg)^5 = ((bcdcd)^5e)^3 \rangle$$

(see [Soi85]), in which

$$G_2(4) \cong \langle a, b, c, d, e, f \rangle,$$

and

$$3_2 \cdot U_4(3) : 2 \cong \langle a, b, c, d, f, g \rangle.$$

A presentation for $Suz : 2$ is obtained from this presentation for Suz by adjoining a generator t, and the relations

$$1 = t^2 = (at)^2 = (bt)^2 = (ct)^2 = (dt)^2 = (et)^2 = f^t g.$$

Note that t normalizes $\langle a, b, c, d, f, g \rangle$.

4.21.1 rank 2

There are no pseudo-permutation characters of rank 2 for G.

4.21.2 rank 3

(i) $\pi = \chi_1 + \chi_4 + \chi_5$ of degree $1782 = 1 + 780 + 1001$.

There is a unique class of subgroups of G of index 1782. A representative is a maximal $H \cong G_2(4)$, and the associated permutation character is π.

The subdegrees are 1, 416, 1365, and the collapsed adjacency matrices are

$$\begin{pmatrix} 0 & 416 & 0 \\ 1 & 100 & 315 \\ 0 & 96 & 320 \end{pmatrix}$$

$$\begin{pmatrix} 0 & 0 & 1365 \\ 0 & 315 & 1050 \\ 1 & 320 & 1044 \end{pmatrix}.$$

The orbital graph corresponding to the suborbit of length 416 is the *Suzuki graph* $\Gamma(Suz)$, which was used by Suzuki [Suz69] in the construction of his sporadic simple group. The automorphism group of $\Gamma(Suz)$ is Aut $(Suz) = Suz : 2$.

4.21.3 rank 4

There are no pseudo-permutation characters of rank 4 for G.

4.21.4 rank 5

(*ii*) $\pi = \chi_1 + \chi_2 + \chi_4 + \chi_6 + \chi_9$ of degree $10296 = 1 + 143 + 780 + 3432 + 5940$.

G has no subgroup of index 10296, and hence π is not a permutation character.

(*iii*) $\pi = \chi_1 + \chi_3 + \chi_4 + \chi_9 + \chi_{15}$ of degree $22880 = 1 + 364 + 780 + 5940 + 15795$.

There is a unique class of subgroups of G of index 22880. A representative is a maximal $H \cong 3_2{\cdot}U_4(3){:}2$, and the associated permutation character is π.

The subdegrees are 1, 280, 486, 8505, 13608, and the collapsed adjacency matrices are

$$\begin{pmatrix} 0 & 280 & 0 & 0 & 0 \\ 1 & 36 & 0 & 243 & 0 \\ 0 & 0 & 0 & 0 & 280 \\ 0 & 8 & 0 & 128 & 144 \\ 0 & 0 & 10 & 90 & 180 \end{pmatrix}$$

$$\begin{pmatrix} 0 & 0 & 486 & 0 & 0 \\ 0 & 0 & 0 & 0 & 486 \\ 1 & 0 & 58 & 315 & 112 \\ 0 & 0 & 18 & 180 & 288 \\ 0 & 10 & 4 & 180 & 292 \end{pmatrix}$$

$$\begin{pmatrix} 0 & 0 & 0 & 8505 & 0 \\ 0 & 243 & 0 & 3888 & 4374 \\ 0 & 0 & 315 & 3150 & 5040 \\ 1 & 128 & 180 & 3300 & 4896 \\ 0 & 90 & 180 & 3060 & 5175 \end{pmatrix}$$

$$\begin{pmatrix} 0 & 0 & 0 & 0 & 13608 \\ 0 & 0 & 486 & 4374 & 8748 \\ 0 & 280 & 112 & 5040 & 8176 \\ 0 & 144 & 288 & 4896 & 8280 \\ 1 & 180 & 292 & 5175 & 7960 \end{pmatrix}.$$

The only non-complete distance-regular generalized orbital graph is

$\Gamma\{2\}$, $\iota = \{280, 243, 144, 10; 1, 8, 90, 280\}$, G acts distance-transitively.

We may take the vertex-set of $\Gamma\{2\}$ to be the conjugacy class in Suz of $3A$-generated subgroups of order 3. Then vertices A, B are joined by an edge in $\Gamma\{2\}$ if and only if $\langle A, B \rangle \cong 3^2$. This graph is known as the *Patterson graph*, and is described in more detail in [BCN89, pp. 410–412].

4.22 Automorphism group $Suz\!:\!2$ of Suz

A presentation for $Suz\!:\!2$ is given in the section for Suz.

4.22.1 rank 2

There are no faithful pseudo-permutation characters of rank 2 for G.

4.22.2 rank 3

(i) $\pi = \chi_1 + \chi_7 + \chi_9$ of degree $1782 = 1 + 780 + 1001$.

There is a unique class of subgroups of G of index 1782. A representative is a maximal $H \cong G_2(4)\!:\!2$, and the associated permutation character is π. The restriction of this representation to Suz is the representation in case (i) for Suz, and so the collapsed adjacency matrices are the same as for that representation.

4.22.3 rank 4

There are no pseudo-permutation characters of rank 4 for G.

4.22.4 rank 5

(ii) $\pi = \chi_1 + \chi_5 + \chi_7 + \chi_{14} + \chi_{23}$ of degree $22880 = 1 + 364 + 780 + 5940 + 15795$.

There is a unique class of subgroups of G of index 22880. A representative is a maximal $H \cong 3_2\!\cdot\! U_4(3)\!:\!(2^2)_{133}$, and the associated permutation character is π. The restriction of this representation to Suz is the representation in case (iii) for Suz, and so the collapsed adjacency matrices and associated non-complete distance-regular graph (on which G acts distance-transitively) are the same as for that representation.

4.23 The O'Nan group $O'N$

In [Soi90], the following presentation is deduced for the O'Nan group $O'N$:

$$\langle g, a3b3c8d3e3f \mid af = g^2 = (cd)^4, 1 = cc^{dgdg} = dd^{cgcg},$$

$$1 = (bcdg)^5 = gbg^c g^b g^c = geg^d g^e g^d \rangle.$$

Note that this presentation contains no Coxeter graph relations at all for the generator g, which has order 4. In this $O'N$, we have

$$L_3(7):2 \cong \langle a, b, c, d, e, f \rangle.$$

Words generating certain other subgroups are given in [Soi90].

4.23.1 rank 2

G has no pseudo-permutation character of rank 2.

4.23.2 rank 3

G has no pseudo-permutation character of rank 3.

4.23.3 rank 4

G has no pseudo-permutation character of rank 4.

4.23.4 rank 5

(i) $\pi = \chi_1 + \chi_2 + \chi_7 + \chi_i + \chi_{11}$ of degree $122760 = 1 + 10944 + 26752 + 32395 + 52668$, where $i = 8$ or 9.

An examination of the table of maximal subgroups of G shows that H lies in one of the two classes of maximal subgroups of index 122760, and $H \cong L_3(7):2$. These classes correspond to the two possibilities for π above, and are interchanged by an outer automorphism of G (see ATLAS).

For each of these two equivalent, but not permutationally isomorphic, representations, the subdegrees are 1, 5586, 6384, 52136, 58653, and the collapsed adjacency matrices are

$$
\begin{pmatrix}
0 & 5586 & 0 & 0 & 0 \\
1 & 364 & 216 & 2464 & 2541 \\
0 & 189 & 301 & 2303 & 2793 \\
0 & 264 & 282 & 2394 & 2646 \\
0 & 242 & 304 & 2352 & 2688
\end{pmatrix}
$$

$$
\begin{pmatrix}
0 & 0 & 6384 & 0 & 0 \\
0 & 216 & 344 & 2632 & 3192 \\
1 & 301 & 349 & 2793 & 2940 \\
0 & 282 & 342 & 2736 & 3024 \\
0 & 304 & 320 & 2688 & 3072
\end{pmatrix}
$$

$$
\begin{pmatrix}
0 & 0 & 0 & 52136 & 0 \\
0 & 2464 & 2632 & 22344 & 24696 \\
0 & 2303 & 2793 & 22344 & 24696 \\
1 & 2394 & 2736 & 22057 & 24948 \\
0 & 2352 & 2688 & 22176 & 24920
\end{pmatrix}
$$

$$
\begin{pmatrix}
0 & 0 & 0 & 0 & 58653 \\
0 & 2541 & 3192 & 24696 & 28224 \\
0 & 2793 & 2940 & 24696 & 28224 \\
0 & 2646 & 3024 & 24948 & 28035 \\
1 & 2688 & 3072 & 24920 & 27972
\end{pmatrix}.
$$

The non-complete distance-regular generalized orbital graphs are

$$\Gamma\{2,3\}, \quad \iota = \{11970, 10829; 1, 1170\} \quad \text{(and complement)}.$$

We believe these distance-regular graphs are new. Let $\Gamma = \Gamma\{2,3\}$. We have $O'N$ acting vertex-primitively on Γ. From [LPS90, Chapter 9, Tables I–VI], we deduce that $soc(\mathrm{Aut}\,(\Gamma)) \cong O'N$. Since a permutation representation of degree 122760 for $O'N$ does not extend to $\mathrm{Aut}\,(O'N) = O'N\!:\!2$, it follows that $\mathrm{Aut}\,(\Gamma) \cong O'N$.

4.24 Automorphism group $O'N{:}2$ of $O'N$

G has no faithful pseudo-permutation character of rank less than or equal to 5.

4.25 The Conway group Co_3

A presentation for Co_3 is

$$\langle b4h3a3b5c3d3f4c3e4a, e6f6h \mid a = (cf)^2, b = (ef)^3,$$

$$d = (bh)^2 = (eah)^3, 1 = (eab)^3 = (bce)^5 = (adfh)^3 = (cef)^7\rangle$$

(see [Soi85, Soi88]), in which

$$McL{:}2 \cong \langle a, b, c, d, e, f, hfdcecbahfecdfehfh\rangle > \langle a, b, c, d, e, f\rangle \cong McL,$$

and

$$HS \cong \langle a, b, c, d, e, h\rangle.$$

We record here that

$$2{\cdot}S_6(2) \cong \langle b, d, e, f, c^{ba}, cbdb^ch, e^ab^cbec, (deacbacde)^{bca}\rangle \text{ (rank 7),}$$

and $M_{23} \cong \langle a, b, c, d, e, (feadfehfhefd)^{cef}\rangle$ (rank 8).

4.25.1 rank 2

(i) $\pi = \chi_1 + \chi_5$ of degree $276 = 1 + 275$.

This character corresponds to the unique 2-transitive representation of G. A point stabilizer is $H \cong McL{:}2$.

4.25.2 rank 3

There are no pseudo-permutation characters of rank 3 for G.

4.25.3 rank 4

(ii) $\pi = \chi_1 + \chi_2 + \chi_4 + \chi_5$ of degree $552 = 1 + 23 + 253 + 275$.

The group G has no maximal subgroup of index 552, and hence $H < L < G$ with $m := |G : L|$ dividing 552. Then $m = 276$ and H has index 2 in $L \cong McL\!:\!2$. Hence $H \cong McL$. Thus G preserves a block system Σ consisting of 276 blocks of size 2, and the action of G on Σ is permutationally isomorphic to the 2-transitive representation in (i) above. Let B be the block of Σ containing α. Then H fixes both points of B, and by the ATLAS, p.100, H is transitive on $\Sigma \setminus \{B\}$. Thus G has rank 3 or 4 on Ω, and, as G has no rank 3 permutation characters, it follows that G has rank 4, subdegrees 1, 1, 275, 275, and associated permutation character π.

The collapsed adjacency matrices are

$$\begin{pmatrix} 0 & 1 & 0 & 0 \\ 1 & 0 & 0 & 0 \\ 0 & 0 & 0 & 1 \\ 0 & 0 & 1 & 0 \end{pmatrix}$$

$$\begin{pmatrix} 0 & 0 & 275 & 0 \\ 0 & 0 & 0 & 275 \\ 1 & 0 & 162 & 112 \\ 0 & 1 & 112 & 162 \end{pmatrix}$$

$$\begin{pmatrix} 0 & 0 & 0 & 275 \\ 0 & 0 & 275 & 0 \\ 0 & 1 & 112 & 162 \\ 1 & 0 & 162 & 112 \end{pmatrix}.$$

The non-complete distance-regular generalized orbital graphs are

$\Gamma\{3\}$, $\iota = \{275, 112, 1; 1, 112, 275\}$, G acts distance-transitively
$\Gamma\{4\}$, $\iota = \{275, 162, 1; 1, 162, 275\}$, G acts distance-transitively
$\Gamma\{3,4\} \cong K_{276\times 2}$.

$\Gamma\{3\}$ and $\Gamma\{4\}$ are well-known Taylor graphs, and are described in [BCN89, p.373].

4.25.4 rank 5

(iii) $\pi = \chi_1 + \chi_2 + \chi_5 + \chi_9 + \chi_{15}$ of degree $11178 = 1 + 23 + 275 + 2024 + 8855$.

There is a unique class of subgroups of G of index 11178. A representative is a maximal $H \cong HS$.

Direct calculation shows that G acting on the cosets of H has rank 5, and so must have permutation character π, the unique pseudo-permutation character of G of rank 5. The subdegrees are 1, 352, 1100, 4125, 5600, and the collapsed adjacency matrices are

$$\begin{pmatrix} 0 & 352 & 0 & 0 & 0 \\ 1 & 1 & 175 & 0 & 175 \\ 0 & 56 & 0 & 240 & 56 \\ 0 & 0 & 64 & 64 & 224 \\ 0 & 11 & 11 & 165 & 165 \end{pmatrix}$$

$$\begin{pmatrix} 0 & 0 & 1100 & 0 & 0 \\ 0 & 175 & 0 & 750 & 175 \\ 1 & 0 & 322 & 105 & 672 \\ 0 & 64 & 28 & 560 & 448 \\ 0 & 11 & 132 & 330 & 627 \end{pmatrix}$$

$$\begin{pmatrix} 0 & 0 & 0 & 4125 & 0 \\ 0 & 0 & 750 & 750 & 2625 \\ 0 & 240 & 105 & 2100 & 1680 \\ 1 & 64 & 560 & 1260 & 2240 \\ 0 & 165 & 330 & 1650 & 1980 \end{pmatrix}$$

$$\begin{pmatrix} 0 & 0 & 0 & 0 & 5600 \\ 0 & 175 & 175 & 2625 & 2625 \\ 0 & 56 & 672 & 1680 & 3192 \\ 0 & 224 & 448 & 2240 & 2688 \\ 1 & 165 & 627 & 1980 & 2827 \end{pmatrix}.$$

There are no non-complete distance-regular generalized orbital graphs.

4.26 The Conway group Co_2

A presentation for Co_2 is

$$\langle a3b5c3d3f6e, a4e3c4f4g4b \mid a = (cf)^2, e = (bg)^2, b = (ef)^3,$$

$$1 = (aecd)^4 = (baefg)^3 = (cef)^7 \rangle$$

(see [Soi85, Soi88]), in which

$$U_6(2):2 \;\cong\; \langle g, cef, dcfd, deaefecdfgfdecdfgbfcde \rangle$$
$$>\; \langle g, cef, dcfd \rangle \cong U_6(2),$$

$$2^{10}: M_{22}:2 \cong \langle a, b, c, d, e, g, (gfdc)^4 \rangle,$$

and

$$2^{1+8}: S_6(2) \cong \langle a, b, d, e, f, g, (gfdc)^4, (abcdefg)^5 \rangle.$$

The short words generating $U_6(2)$ were found by M. Schönert.

We record here that $McL \cong \langle a, b, c, d, e, f \rangle$ (rank 6), and $HS:2 \cong \langle a, b, c, d, f, (adecgbcdeabc)^{bce} \rangle$ (rank 7).

4.26.1 rank 2

There are no pseudo-permutation characters of rank 2 for G.

4.26.2 rank 3

(i) $\pi = \chi_1 + \chi_4 + \chi_6$ of degree $2300 = 1 + 275 + 2024$.

There is a unique class of subgroups of G of index 2300. A representative is a maximal $H \cong U_6(2):2$, and the associated permutation character is π.

The subdegrees are 1, 891, 1408, and the collapsed adjacency matrices are

$$\begin{pmatrix} 0 & 891 & 0 \\ 1 & 378 & 512 \\ 0 & 324 & 567 \end{pmatrix}$$

$$\begin{pmatrix} 0 & 0 & 1408 \\ 0 & 512 & 896 \\ 1 & 567 & 840 \end{pmatrix}.$$

4.26.3 rank 4

There are no pseudo-permutation characters of rank 4 for G.

4.26.4 rank 5

(ii) $\pi = \chi_1 + \chi_2 + \chi_4 + \chi_6 + \chi_7$ of degree $4600 = 1 + 23 + 275 + 2024 + 2277$.

The group G has no maximal subgroups of index 4600, and hence $H < L < G$ with $m := |G : L|$ dividing 4600. Then $m = 2300$ and H has index 2 in $L \cong U_6(2){:}2$. Hence $H \cong U_6(2)$. Thus G preserves a block system Σ consisting of 2300 blocks of size 2, and the action of G on Σ is permutationally isomorphic to the rank 3 representation in (i) above with subdegrees 1, 891, 1408. Let Σ_1, Σ_2 denote the orbits of L on Σ of lengths 891, 1408 respectively. Let $B \in \Sigma$ be the block containing α, and let $B' \in \Sigma_1$. Since $|L : H| = 2$ is prime to 891, H is transitive on Σ_1 and so $L_{B'}$ is transitive on B. By the ATLAS, p.115, $L_{B'} \cong 2^9{:}L_3(4){:}2$, which has a unique subgroup of index 2, namely $H_{B'} \cong 2^9{:}L_3(4)$. Hence, since $H_{B'}$ has no subgroup of index 2, $H_{B'}$ fixes B' pointwise, and so H has two orbits of length 891 on the points of Ω in the blocks of Σ_1.

Now by the ATLAS, p.115, L has a unique class of subgroups of index 1408, so for $B'' \in \Sigma_2$, $L_{B''} \cong U_4(3).2^2$. As $H \cong U_6(2)$ has no subgroup of index 704, H is transitive on Σ_2 and $H_{B''} \cong U_4(3){:}2$. Thus H is either transitive on the 2816 points of Ω contained in Σ_2, or has two orbits of length 1408 in this set of points. Suppose the latter is true. Then the representation of G on the cosets of H has rank 6, and is imprimitive with 2300 blocks of size 2; the permutation character π' for the action of G on the cosets of H must contain the character corresponding to the action of G on this system of imprimitivity, that is, π' contains $\chi_1 + \chi_4 + \chi_6$, and there are three other irreducible characters of G whose degrees sum to 2300. Checking the character table for G in the ATLAS shows that this is not possible. Hence H is transitive on these 2816 points, G has rank 5, and the subdegrees are 1, 1, 891, 891, 2816.

Thus we have described, up to permutational isomorphism, a unique imprimitive rank 5 representation of G. The collapsed adjacency matrices are

$$\begin{pmatrix} 0 & 1 & 0 & 0 & 0 \\ 1 & 0 & 0 & 0 & 0 \\ 0 & 0 & 0 & 1 & 0 \\ 0 & 0 & 1 & 0 & 0 \\ 0 & 0 & 0 & 0 & 1 \end{pmatrix}$$

$$\begin{pmatrix} 0 & 0 & 891 & 0 & 0 \\ 0 & 0 & 0 & 891 & 0 \\ 1 & 0 & 336 & 42 & 512 \\ 0 & 1 & 42 & 336 & 512 \\ 0 & 0 & 162 & 162 & 567 \end{pmatrix}$$

$$\begin{pmatrix} 0 & 0 & 0 & 891 & 0 \\ 0 & 0 & 891 & 0 & 0 \\ 0 & 1 & 42 & 336 & 512 \\ 1 & 0 & 336 & 42 & 512 \\ 0 & 0 & 162 & 162 & 567 \end{pmatrix}$$

$$\begin{pmatrix} 0 & 0 & 0 & 0 & 2816 \\ 0 & 0 & 0 & 0 & 2816 \\ 0 & 0 & 512 & 512 & 1792 \\ 0 & 0 & 512 & 512 & 1792 \\ 1 & 1 & 567 & 567 & 1680 \end{pmatrix}.$$

The only non-complete distance-regular generalized orbital graph is

$$\Gamma\{3,4,5\} \cong K_{2300\times 2}.$$

(iii) $\pi = \chi_1 + \chi_4 + \chi_6 + \chi_{14} + \chi_{17}$ of degree $46575 = 1 + 275 + 2024 + 12650 + 31625.$

There is a unique class of subgroups of G of index 46575. A representative is a maximal $H \cong 2^{10}{:}M_{22}{:}2$.

Direct calculation shows that G acting on the cosets of H has rank 5, and so must have permutation character π, the unique rank 5 pseudo-permutation character of G of degree 46575. The subdegrees are 1, 462,

2464, 21120, 22528, and the collapsed adjacency matrices are

$$\begin{pmatrix} 0 & 462 & 0 & 0 & 0 \\ 1 & 61 & 80 & 320 & 0 \\ 0 & 15 & 15 & 240 & 192 \\ 0 & 7 & 28 & 203 & 224 \\ 0 & 0 & 21 & 210 & 231 \end{pmatrix}$$

$$\begin{pmatrix} 0 & 0 & 2464 & 0 & 0 \\ 0 & 80 & 80 & 1280 & 1024 \\ 1 & 15 & 496 & 480 & 1472 \\ 0 & 28 & 56 & 1260 & 1120 \\ 0 & 21 & 161 & 1050 & 1232 \end{pmatrix}$$

$$\begin{pmatrix} 0 & 0 & 0 & 21120 & 0 \\ 0 & 320 & 1280 & 9280 & 10240 \\ 0 & 240 & 480 & 10800 & 9600 \\ 1 & 203 & 1260 & 9352 & 10304 \\ 0 & 210 & 1050 & 9660 & 10200 \end{pmatrix}$$

$$\begin{pmatrix} 0 & 0 & 0 & 0 & 22528 \\ 0 & 0 & 1024 & 10240 & 11264 \\ 0 & 192 & 1472 & 9600 & 11264 \\ 0 & 224 & 1120 & 10304 & 10880 \\ 1 & 231 & 1232 & 10200 & 10864 \end{pmatrix}.$$

There are no non-complete distance-regular generalized orbital graphs.

(*iv*) $\pi = \chi_1 + \chi_4 + \chi_6 + \chi_{15} + \chi_{17}$ of degree $56925 = 1 + 275 + 2024 + 23000 + 31625$.

There is a unique class of subgroups of G of index 56925. A representative is a maximal $H \cong 2^{1+8}_+ : S_6(2)$.

Direct calculation shows that G acting on the cosets of H has rank 5, and so must have permutation character π, the unique rank 5 pseudo-permutation character of G of degree 56925. The subdegrees are 1, 1008, 1260, 14336, 40320, and the collapsed adjacency matrices are

$$\begin{pmatrix} 0 & 1008 & 0 & 0 & 0 \\ 1 & 0 & 135 & 512 & 360 \\ 0 & 108 & 36 & 0 & 864 \\ 0 & 36 & 0 & 162 & 810 \\ 0 & 9 & 27 & 288 & 684 \end{pmatrix}$$

$$\begin{pmatrix} 0 & 0 & 1260 & 0 & 0 \\ 0 & 135 & 45 & 0 & 1080 \\ 1 & 36 & 135 & 512 & 576 \\ 0 & 0 & 45 & 405 & 810 \\ 0 & 27 & 18 & 288 & 927 \end{pmatrix}$$

$$\begin{pmatrix} 0 & 0 & 0 & 14336 & 0 \\ 0 & 512 & 0 & 2304 & 11520 \\ 0 & 0 & 512 & 4608 & 9216 \\ 1 & 162 & 405 & 4048 & 9720 \\ 0 & 288 & 288 & 3456 & 10304 \end{pmatrix}$$

$$\begin{pmatrix} 0 & 0 & 0 & 0 & 40320 \\ 0 & 360 & 1080 & 11520 & 27360 \\ 0 & 864 & 576 & 9216 & 29664 \\ 0 & 810 & 810 & 9720 & 28980 \\ 1 & 684 & 927 & 10304 & 28404 \end{pmatrix}.$$

The non-complete distance-regular generalized orbital graphs are

$$\Gamma\{2,3\}, \quad \iota = \{2268, 1952; 1, 81\} \quad \text{(and complement)}.$$

We believe these distance-regular graphs are new. Let $\Gamma = \Gamma\{2,3\}$. We may take the vertex-set of Γ to be the conjugacy class $2A$ of Co_2. Then structure constant calculations (see [Wil86]) show that, in Γ, two vertices x, y are joined by an edge if and only if xy has order 2. We have Co_2 acting vertex-primitively on Γ. From [LPS90, Chapter 9, Tables I–VI], we deduce that $soc(\text{Aut}\,(\Gamma)) \cong Co_2$. Since Co_2 has a trivial outer automorphism group, it follows that $\text{Aut}\,(\Gamma) \cong Co_2$.

4.27 The Fischer group Fi_{22}

A presentation for Fi_{22} is

$$\langle a3b3c3d3e3f3g, d3h3i \mid 1 = (dcbdefdhi)^{10} = (abcdefh)^9 = (bcdefgh)^9 \rangle$$

(this is $Y_{332}/Z(Y_{332})$; see [CNS88]), in which

$$2 \cdot U_6(2) \cong \langle a, c, d, e, f, g, h, i, (abcdeh)^5 \rangle,$$

and

$$O_7(3) \cong \langle b, c, d, e, f, g, h, i \rangle.$$

We record here that

$$2^{10}\colon M_{22} \cong \langle a, c, e, g, h, bacb, dced, fegf, dchd, dehd, (cdehi)^4 \rangle \text{ (rank 8)},$$

and $2^6\colon S_6(2) \cong \langle a, b, c, d, e, f, g, h \rangle$ (rank 10).

A presentation for $Fi_{22}\colon 2$ can be obtained from that of Fi_{22} by adjoining the generator t, and the relations

$$1 = t^2 = a^t g = b^t f = c^t e = (dt)^2 = (ht)^2 = (it)^2.$$

In this $Fi_{22}\colon 2$, we have that

$$O_8^+(2)\colon S_3 \times 2 \cong C(t) = \langle d, h, i, ag, bf, ce, t \rangle.$$

4.27.1 rank 2

G has no pseudo-permutation character of rank 2.

4.27.2 rank 3

(i) $\pi = \chi_1 + \chi_3 + \chi_7$ of degree $3510 = 1 + 429 + 3080$.

An examination of the table of maximal subgroups of G shows that H is maximal in G and $H \cong 2{\cdot}U_6(2)$. The subdegrees are 1, 693, 2816, and the collapsed adjacency matrices are

$$\begin{pmatrix} 0 & 693 & 0 \\ 1 & 180 & 512 \\ 0 & 126 & 567 \end{pmatrix}$$

$$\begin{pmatrix} 0 & 0 & 2816 \\ 0 & 512 & 2304 \\ 1 & 567 & 2248 \end{pmatrix}.$$

The vertex-set of these orbital graphs may be taken to be the class of *3-transpositions* of G (that is, a conjugacy class D of involutions of G, such that, for all $d, e \in D$, de has order 1, 2, or 3). For the first orbital graph we join d, e exactly when de has order 2 (equivalently $d \neq e$ and d, e commute), and for the second graph we join d, e exactly when de has order 3. These graphs were originally studied by Fischer [Fis69].

(ii) $\pi = \chi_1 + \chi_3 + \chi_9$ of degree $14080 = 1 + 429 + 13650$.

An examination of the table of maximal subgroups of G shows that H lies in one of two classes of maximal subgroups of index 14080, $H \cong O_7(3)$. These representations are interchanged by an outer automorphism of G.

Thus we have described two equivalent, but not permutationally isomorphic, primitive rank 3 representations of G. The subdegrees are 1, 3159, 10920, and the collapsed adjacency matrices are

$$\begin{pmatrix} 0 & 3159 & 0 \\ 1 & 918 & 2240 \\ 0 & 648 & 2511 \end{pmatrix}$$

$$\begin{pmatrix} 0 & 0 & 10920 \\ 0 & 2240 & 8680 \\ 1 & 2511 & 8408 \end{pmatrix}.$$

Remark The group $\mathrm{Aut}\,(G) = G{:}2$ acts with permutation rank 6 on the cosets of $H < G$. This imprimitive representation has subdegrees 1, 364, 1080, 3159, 10920, 12636, and the generalized orbital digraph $\Gamma\{2,3\}$ for this representation is distance-regular, with intersection array $\{1444, 1443, 1296; 1, 148, 1444\}$. This graph is the incidence graph of a square $2 - (20160, 1444, 148)$ design, on which Fi_{22} acts as a group of automorphisms. We have been informed by P.J. Cameron that this design for Fi_{22} was first discovered by A. Rudvalis in the early 1970s.

4.27.3 rank 4

(*iii*) $\pi = \chi_1 + \chi_3 + \chi_7 + \chi_9$ of degree $17160 = 1 + 429 + 3080 + 13650$.

G has no subgroup of index 17160.

(*iv*) $\pi = \chi_1 + \chi_7 + \chi_9 + \chi_{13}$ of degree $61776 = 1 + 3080 + 13650 + 45045$.

An examination of the indices of maximal subgroups of G shows that H is maximal in G and $H \cong O_8^+(2){:}S_3$. The subdegrees are 1, 1575, 22400, 37800 and the collapsed adjacency matrices are

$$\begin{pmatrix} 0 & 1575 & 0 & 0 \\ 1 & 198 & 512 & 864 \\ 0 & 36 & 567 & 972 \\ 0 & 36 & 576 & 963 \end{pmatrix}$$

$$\begin{pmatrix} 0 & 0 & 22400 & 0 \\ 0 & 512 & 8064 & 13824 \\ 1 & 567 & 8224 & 13608 \\ 0 & 576 & 8064 & 13760 \end{pmatrix}$$

$$\begin{pmatrix} 0 & 0 & 0 & 37800 \\ 0 & 864 & 13824 & 23112 \\ 0 & 972 & 13608 & 23220 \\ 1 & 963 & 13760 & 23076 \end{pmatrix}.$$

The non-complete distance-regular generalized orbital graphs are

$$\Gamma\{2\}, \quad \iota = \{1575, 1376; 1, 36\} \qquad \text{(and complement)}$$
$$\Gamma\{3\}, \quad \iota = \{22400, 14175; 1, 8064\} \quad \text{(and complement)}.$$

We believe these distance-regular graphs are new. The vertex-sets of $\Gamma\{2\}$ and $\Gamma\{3\}$ may be taken to be the conjugacy class $2D$ of $Fi_{22}:2$. Then structure constant calculations (see [Wil86]) show that vertices x, y are joined by an edge in $\Gamma\{2\}$ (respectively $\Gamma\{3\}$) if and only if xy has order 2 (respectively 3). Let $\Gamma = \Gamma\{2\}$ or $\Gamma\{3\}$. Then we have Fi_{22} acting vertex-primitively on Γ. From [LPS90, Chapter 9, Tables I–VI], we deduce that $soc(\text{Aut}\,(\Gamma)) \cong Fi_{22}$. Since the permutation representation in the present case extends to a representation of the same rank for $\text{Aut}\,(Fi_{22}) = Fi_{22}:2$, it follows that $\text{Aut}\,(\Gamma) \cong Fi_{22}:2$.

4.27.4 rank 5

G has no pseudo-permutation character of rank 5.

4.28 Automorphism group $Fi_{22}:2$ of Fi_{22}

A presentation for $Fi_{22}:2$ is given in the section for Fi_{22}.

4.28.1 rank 2

G has no faithful pseudo-permutation character of rank 2.

4.28.2 rank 3

(*i*) $\pi = \chi_1 + \chi_5 + \chi_{13}$ of degree $3510 = 1 + 429 + 3080$.

There is a unique class of subgroups of G of index 3510. A representative is a maximal $H \cong 2 \cdot U_6(2).2$, and the associated permutation character is π. The restriction of this representation to Fi_{22} is the representation in case (*i*) for Fi_{22}, and so the collapsed adjacency matrices are the same as for that representation.

4.28.3 rank 4

(*ii*) $\pi = \chi_1 + \chi_{13} + \chi_{17} + \chi_{25}$ of degree $61776 = 1 + 3080 + 13650 + 45045$.

There is a unique class of subgroups of G of index 61776. A representative is a maximal $H \cong O_8^+(2){:}S_3 \times 2$, and the associated permutation character is π. The restriction of this representation to Fi_{22} is the representation in case (*iv*) for Fi_{22}, and so the collapsed adjacency matrices and associated distance-regular graphs are the same as for that representation.

4.28.4 rank 5

G has no pseudo-permutation character of rank 5.

4.29 The Harada-Norton group HN

G has no faithful pseudo-permutation character of rank less than or equal to 5.

4.30 Automorphism group $HN{:}2$ of HN

G has no faithful pseudo-permutation character of rank less than or equal to 5.

4.31 The Lyons group *Ly*

C.C. Sims derived a presentation for the Lyons group *Ly*, as part of his proof of the existence and uniqueness of this group (see [Sim73]). The usefulness of the presentation as given in [Sim73] depends on an unpublished presentation of $G_2(5)$, which now appears in [JW96].

4.31.1 rank 2

G has no pseudo-permutation character of rank 2.

4.31.2 rank 3

G has no pseudo-permutation character of rank 3.

4.31.3 rank 4

G has no pseudo-permutation character of rank 4.

4.31.4 rank 5

(i) $\pi = \chi_1 + \chi_4 + \chi_{11} + \chi_{12} + \chi_{14}$ of degree $8835156 = 1 + 45694 + 1534500 + 3028266 + 4226695$.

An examination of the table of maximal subgroups of G shows that H is maximal in G, and $H \cong G_2(5)$. The subdegrees are

$$1, 19530, 968750, 2034375, 5812500.$$

W.M. Kantor [Kan81] gives the collapsed adjacency matrix A for the orbital digraph corresponding to the suborbit of length 19530. The matrix A has distinct eigenvalues, allowing the collapsed adjacency matrices for the other orbital digraphs of this multiplicity-free representation to be determined from A (this uses some basic theory of Bose-Mesner algebras; see [BCN89, pp. 43–46]). This was done by D.V. Pasechnik, using a *Maple* program he wrote for this purpose.

The collapsed adjacency matrices are

$$
\begin{pmatrix}
0 & 19530 & 0 & 0 & 0 \\
1 & 154 & 3125 & 3750 & 12500 \\
0 & 63 & 2520 & 4599 & 12348 \\
0 & 36 & 2190 & 4584 & 12720 \\
0 & 42 & 2058 & 4452 & 12978
\end{pmatrix}
$$

$$
\begin{pmatrix}
0 & 0 & 968750 & 0 & 0 \\
0 & 3125 & 125000 & 228125 & 612500 \\
1 & 2520 & 114013 & 225372 & 626844 \\
0 & 2190 & 107320 & 223080 & 636160 \\
0 & 2058 & 104474 & 222656 & 639562
\end{pmatrix}
$$

$$
\begin{pmatrix}
0 & 0 & 0 & 2034375 & 0 \\
0 & 3750 & 228125 & 477500 & 1325000 \\
0 & 4599 & 225372 & 468468 & 1335936 \\
1 & 4584 & 223080 & 468670 & 1338040 \\
0 & 4452 & 222656 & 468314 & 1338953
\end{pmatrix}
$$

$$
\begin{pmatrix}
0 & 0 & 0 & 0 & 5812500 \\
0 & 12500 & 612500 & 1325000 & 3862500 \\
0 & 12348 & 626844 & 1335936 & 3837372 \\
0 & 12720 & 636160 & 1338040 & 3825580 \\
1 & 12978 & 639562 & 1338953 & 3821006
\end{pmatrix}.
$$

There are no non-complete distance-regular generalized orbital graphs.

(*ii*) $\pi = \chi_1 + \chi_4 + \chi_{11} + \chi_{12} + \chi_{15}$ of degree $9606125 = 1 + 45694 + 1534500 + 3028266 + 4997664$.

An examination of the table of maximal subgroups of G shows that H is maximal in G, and $H \cong 3 \cdot McL : 2$. The subdegrees are

$$1, 15400, 534600, 1871100, 7185024.$$

Cooperman, Finkelstein, York and Tselman [CFYT94] have constructed this permutation representation, using a 111-dimensional $GF(5)$ matrix representation for Ly constructed by Meyer, Neutsch and Parker [MNP85] (see also [Gol95]). We used this explicit permutation representation to construct the collapsed adjacency matrices below on an IBM RS/6000 of the Institute for Experimental Mathematics, University of Essen.

The collapsed adjacency matrices are

$$
\begin{pmatrix}
0 & 15400 & 0 & 0 & 0 \\
1 & 90 & 1215 & 2430 & 11664 \\
0 & 35 & 560 & 3045 & 11760 \\
0 & 20 & 870 & 2990 & 11520 \\
0 & 25 & 875 & 3000 & 11500
\end{pmatrix}
$$

$$
\begin{pmatrix}
0 & 0 & 534600 & 0 & 0 \\
0 & 1215 & 19440 & 105705 & 408240 \\
1 & 560 & 38985 & 102270 & 392784 \\
0 & 870 & 29220 & 104190 & 400320 \\
0 & 875 & 29225 & 104250 & 400250
\end{pmatrix}
$$

$$
\begin{pmatrix}
0 & 0 & 0 & 1871100 & 0 \\
0 & 2430 & 105705 & 363285 & 1399680 \\
0 & 3045 & 102270 & 364665 & 1401120 \\
1 & 2990 & 104190 & 366255 & 1397664 \\
0 & 3000 & 104250 & 363975 & 1399875
\end{pmatrix}
$$

$$
\begin{pmatrix}
0 & 0 & 0 & 0 & 7185024 \\
0 & 11664 & 408240 & 1399680 & 5365440 \\
0 & 11760 & 392784 & 1401120 & 5379360 \\
0 & 11520 & 400320 & 1397664 & 5375520 \\
1 & 11500 & 400250 & 1399875 & 5373398
\end{pmatrix}.
$$

There are no non-complete distance-regular generalized orbital graphs.

Before the construction of these collapsed adjacency matrices, it was shown [Soi93a] that Ly does not act distance-transitively on any (non-trivial) orbital graph for this representation. The argument made use of the collapsed adjacency matrices in case (v) of McL.

4.32 The Thompson group Th

G has no faithful pseudo-permutation character of rank less than or equal to 5.

4.33 The Fischer group Fi_{23}

A presentation for Fi_{23} is

$$\langle a3b3c3d3e3f3g3j, d3h3i \mid 1 = (dcbdefdhi)^{10} = (abcdefh)^9 \rangle$$

(this is $Y_{342}/Z(Y_{342})$; see [CNS88]), in which

$$2 \cdot Fi_{22} \cong \langle a, b, c, d, e, f, g, h, i \rangle,$$

$$O_8^+(3) : S_3 \cong \langle a, b, c, d, e, f, g, i, j, (abcdeh)^5, a^{(bcdefghi)^5} \rangle$$

which contains

$$\langle ab, ac, ad, ae, af, ag, ai, aj, a(abcdeh)^5, aa^{(bcdefghi)^5} \rangle \cong O_8^+(3) : 3,$$

and we have

$$O_8^+(3) : 2 \cong \langle b, c, d, e, f, g, h, i, j \rangle.$$

4.33.1 rank 2

G has no pseudo-permutation character of rank 2.

4.33.2 rank 3

(i) $\pi = \chi_1 + \chi_2 + \chi_6$ of degree $31671 = 1 + 782 + 30888$.

There is a unique class of subgroups of G of index 31671 (see [KPW89]). A representative is a maximal $H \cong 2 \cdot Fi_{22}$, and the associated permutation character is π. The subdegrees are 1, 3510, 28160, and the collapsed adjacency matrices are

$$\begin{pmatrix} 0 & 3510 & 0 \\ 1 & 693 & 2816 \\ 0 & 351 & 3159 \end{pmatrix}$$

$$\begin{pmatrix} 0 & 0 & 28160 \\ 0 & 2816 & 25344 \\ 1 & 3159 & 25000 \end{pmatrix}.$$

The vertex-set of these orbital graphs may be taken to be the class of 3-*transpositions* of G (recall that this is a conjugacy class D of involutions of G, such that, for all $d, e \in D$, de has order 1, 2, or 3). For the first orbital graph we join d, e exactly when de has order 2 (equivalently $d \neq e$ and d, e commute), and for the second graph we join d, e exactly when de has order 3. These graphs were originally studied by Fischer [Fis69].

(*ii*) $\pi = \chi_1 + \chi_6 + \chi_8$ of degree $137632 = 1 + 30888 + 106743$.

There is a unique class of subgroups of G of index 137632 (see [KPW89]). A representative is a maximal $H \cong O_8^+(3){:}S_3$, and the associated permutation character is π. The subdegrees are 1, 28431, 109200, and the collapsed adjacency matrices are

$$\begin{pmatrix} 0 & 28431 & 0 \\ 1 & 6030 & 22400 \\ 0 & 5832 & 22599 \end{pmatrix}$$

$$\begin{pmatrix} 0 & 0 & 109200 \\ 0 & 22400 & 86800 \\ 1 & 22599 & 86600 \end{pmatrix}.$$

4.33.3 rank 4

G has no pseudo-permutation character of rank 4.

4.33.4 rank 5

(*iii*) $\pi = \chi_1 + \chi_5 + \chi_6 + \chi_8 + \chi_9$ of degree $275264 = 2 \times 137632 = 1 + 25806 + 30888 + 106743 + 111826$.

By [KPW89] there is no maximal subgroup of G of index 275264, and H is the unique subgroup of index 2 in a maximal subgroup $L \cong O_8^+(3){:}S_3$ of G, so $H \cong O_8^+(3){:}3$.

Direct calculation shows that G has rank 5 on the cosets of H, and so the associated permutation character is π, the unique pseudo-permutation character of G of degree 275264. The subdegrees are 1, 1, 28431, 28431, 218400, and the collapsed adjacency matrices are

$$\begin{pmatrix} 0 & 1 & 0 & 0 & 0 \\ 1 & 0 & 0 & 0 & 0 \\ 0 & 0 & 0 & 1 & 0 \\ 0 & 0 & 1 & 0 & 0 \\ 0 & 0 & 0 & 0 & 1 \end{pmatrix}$$

$$\begin{pmatrix} 0 & 0 & 28431 & 0 & 0 \\ 0 & 0 & 0 & 28431 & 0 \\ 1 & 0 & 2880 & 3150 & 22400 \\ 0 & 1 & 3150 & 2880 & 22400 \\ 0 & 0 & 2916 & 2916 & 22599 \end{pmatrix}$$

$$\begin{pmatrix} 0 & 0 & 0 & 28431 & 0 \\ 0 & 0 & 28431 & 0 & 0 \\ 0 & 1 & 3150 & 2880 & 22400 \\ 1 & 0 & 2880 & 3150 & 22400 \\ 0 & 0 & 2916 & 2916 & 22599 \end{pmatrix}$$

$$\begin{pmatrix} 0 & 0 & 0 & 0 & 218400 \\ 0 & 0 & 0 & 0 & 218400 \\ 0 & 0 & 22400 & 22400 & 173600 \\ 0 & 0 & 22400 & 22400 & 173600 \\ 1 & 1 & 22599 & 22599 & 173200 \end{pmatrix}.$$

The only non-complete distance-regular generalized orbital graph is

$$\Gamma\{3,4,5\} \cong K_{137632 \times 2}.$$

(*iv*) $\pi = \chi_1 + \chi_2 + \chi_6 + \chi_8 + \chi_{10}$ of degree $412896 = 3 \times 137632 = 1 + 782 + 30888 + 106743 + 274482$.

By [KPW89] there is no maximal subgroup of G of index 412896, and H is a subgroup of index 3 in a maximal subgroup $L \cong O_8^+(3){:}S_3$ of G, so $H \cong O_8^+(3){:}2$.

Direct calculation shows that G has rank 5 on the cosets of H, and so the associated permutation character is π, the unique pseudo-permutation character of G of degree 412896. The subdegrees are 1, 2, 28431, 56862, 327600, and the collapsed adjacency matrices are

$$\begin{pmatrix} 0 & 2 & 0 & 0 & 0 \\ 1 & 1 & 0 & 0 & 0 \\ 0 & 0 & 0 & 2 & 0 \\ 0 & 0 & 1 & 1 & 0 \\ 0 & 0 & 0 & 0 & 2 \end{pmatrix}$$

$$\begin{pmatrix} 0 & 0 & 28431 & 0 & 0 \\ 0 & 0 & 0 & 28431 & 0 \\ 1 & 0 & 4110 & 1920 & 22400 \\ 0 & 1 & 960 & 5070 & 22400 \\ 0 & 0 & 1944 & 3888 & 22599 \end{pmatrix}$$

$$\begin{pmatrix} 0 & 0 & 0 & 56862 & 0 \\ 0 & 0 & 28431 & 28431 & 0 \\ 0 & 2 & 1920 & 10140 & 44800 \\ 1 & 1 & 5070 & 6990 & 44800 \\ 0 & 0 & 3888 & 7776 & 45198 \end{pmatrix}$$

$$\begin{pmatrix} 0 & 0 & 0 & 0 & 327600 \\ 0 & 0 & 0 & 0 & 327600 \\ 0 & 0 & 22400 & 44800 & 260400 \\ 0 & 0 & 22400 & 44800 & 260400 \\ 1 & 2 & 22599 & 45198 & 259800 \end{pmatrix}.$$

The only non-complete distance-regular generalized orbital graph is

$$\Gamma\{3,4,5\} \cong K_{137632 \times 3}.$$

4.34 The Conway group Co_1

A presentation for $2 \cdot Co_1$ is

$$\langle b4h3a3b5c3d3f4c3e4a, b4g4f, e6f6h \mid a = (cf)^2, e = (bg)^2, b = (ef)^3,$$

$$d = (bh)^2 = (eah)^3, 1 = (adfh)^3 = (baefg)^3 = (cef)^7 \rangle,$$

in which the central involution is

$$z := (adefcefgh)^{39}$$

(see [Soi85, Soi88]), and we have

$$2 \times Co_2 \cong \langle a, b, c, d, e, f, g, z \rangle.$$

We record here that $2^{12} : M_{24} \cong \langle a, b, d, e, f, g, h, (gfdc)^4, z \rangle$ (rank 6), and $2 \times Co_3 \cong \langle a, b, c, d, e, f, h, z \rangle$ (rank 7).

Note that a presentation for Co_1 is obtained by adjoining the relation $z = 1$ to the presentation for $2 \cdot Co_1$ above. However, for many of the subgroups H of Co_1 considered here, to obtain the representation of

Co_1 on the cosets of H by coset enumeration, rather than using this presentation for Co_1, it is more efficient to enumerate the cosets of $2.H$ in $2\cdot Co_1$, using the presentation for $2\cdot Co_1$.

We also make use of the following presentation for $2 \times Co_1$ [Soi87a]:

$$\langle a3b3c8d3e3f3g3h3i \mid a = (cd)^4, 1 = (bcde)^8 \rangle,$$

in which the central involution is

$$z := ((bcdcdefgh)^{13}i)^3,$$

and we have

$$2 \times 3\cdot Suz\colon 2 \cong \langle a, b, c, d, e, f, g, h, z \rangle.$$

As with the presentation for $2\cdot Co_1$, a presentation for Co_1 is obtained by adjoining the relation $z = 1$ to the presentation for $2 \times Co_1$ above, and similar comments apply concerning coset enumeration.

4.34.1 rank 2

There are no pseudo-permutation characters of rank 2 for G.

4.34.2 rank 3

There are no pseudo-permutation characters of rank 3 for G.

4.34.3 rank 4

(i) $\pi = \chi_1 + \chi_3 + \chi_6 + \chi_{10}$ of degree $98280 = 1 + 299 + 17250 + 80730$.

The group G has a unique class of subgroups of index 98280, with representative a maximal $H \cong Co_2$. This gives rise to a primitive rank 4 representation of G, which must have character π, the unique pseudo-permutation character of G of rank 4.

The subdegrees are 1, 4600, 46575, 47104, and the collapsed adjacency matrices are

$$\begin{pmatrix} 0 & 4600 & 0 & 0 \\ 1 & 892 & 891 & 2816 \\ 0 & 88 & 2464 & 2048 \\ 0 & 275 & 2025 & 2300 \end{pmatrix}$$

$$\begin{pmatrix} 0 & 0 & 46575 & 0 \\ 0 & 891 & 24948 & 20736 \\ 1 & 2464 & 21582 & 22528 \\ 0 & 2025 & 22275 & 22275 \end{pmatrix}$$

$$\begin{pmatrix} 0 & 0 & 0 & 47104 \\ 0 & 2816 & 20736 & 23552 \\ 0 & 2048 & 22528 & 22528 \\ 1 & 2300 & 22275 & 22528 \end{pmatrix}.$$

There are no non-complete distance-regular generalized orbital graphs.

4.34.4 rank 5

(ii) $\pi = \chi_1 + \chi_4 + \chi_7 + \chi_{16} + \chi_{20}$ of degree $1545600 = 1 + 1771 + 27300 + 644644 + 871884$.

The group G has a unique class of subgroups of index 1545600. A representative is a maximal $H \cong 3 \cdot Suz : 2$. This gives rise to a primitive rank 5 representation of G, which must have character π, the unique pseudo-permutation character of G of rank 5.

The subdegrees are 1, 5346, 22880, 405405, 1111968 and the collapsed adjacency matrices are

$$\begin{pmatrix} 0 & 5346 & 0 & 0 & 0 \\ 1 & 418 & 0 & 4095 & 832 \\ 0 & 0 & 486 & 0 & 4860 \\ 0 & 54 & 0 & 1836 & 3456 \\ 0 & 4 & 100 & 1260 & 3982 \end{pmatrix}$$

$$\begin{pmatrix} 0 & 0 & 22880 & 0 & 0 \\ 0 & 0 & 2080 & 0 & 20800 \\ 1 & 486 & 280 & 8505 & 13608 \\ 0 & 0 & 480 & 5120 & 17280 \\ 0 & 100 & 280 & 6300 & 16200 \end{pmatrix}$$

$$\begin{pmatrix} 0 & 0 & 0 & 405405 & 0 \\ 0 & 4095 & 0 & 139230 & 262080 \\ 0 & 0 & 8505 & 90720 & 306180 \\ 1 & 1836 & 5120 & 111600 & 286848 \\ 0 & 1260 & 6300 & 104580 & 293265 \end{pmatrix}$$

$$\begin{pmatrix} 0 & 0 & 0 & 0 & 1111968 \\ 0 & 832 & 20800 & 262080 & 828256 \\ 0 & 4860 & 13608 & 306180 & 787320 \\ 0 & 3456 & 17280 & 286848 & 804384 \\ 1 & 3982 & 16200 & 293265 & 798520 \end{pmatrix}.$$

There are no non-complete distance-regular generalized orbital graphs.

4.35 The Janko group J_4

G has no faithful pseudo-permutation character of rank less than or equal to 5.

4.36 The Fischer group Fi'_{24}

A presentation for $3 \cdot Fi_{24}$ is

$$\langle l3k3a3b3c3d3e3f3g3j, d3h3i \mid l = (abcdefh)^9 \rangle$$

(this is isomorphic to Y_{542}; see [CNS88] and [CP92]), in which the normal subgroup of order 3 is generated by

$$z := (dcbakldefgjdhi)^{17},$$

and we have

$$Fi_{23} \times S_3 \cong \langle a, b, c, d, e, f, g, h, i, j, l, z \rangle.$$

Note that a presentation for Fi_{24} is obtained by adjoining the relation $z = 1$ to the presentation for $3 \cdot Fi_{24}$ above. However, to obtain the representation of Fi_{24} on the cosets of $Fi_{23} \times 2$ by coset enumeration, it is best to enumerate the cosets of $Fi_{23} \times S_3$ in the $3 \cdot Fi_{24}$ above.

We do not give a presentation for Fi'_{24}. The representation for $Fi_{24} = Fi'_{24} : 2$ on the cosets of $Fi_{23} \times 2$ restricts to a representation of the same rank for Fi'_{24} on the cosets of Fi_{23}.

4.36.1 rank 2

G has no pseudo-permutation character of rank 2.

4.36.2 rank 3

(i) $\pi = \chi_1 + \chi_3 + \chi_4$ of degree $306936 = 1 + 57477 + 249458$.

By [LW91] there is a unique class of subgroups of G of index 306936. A representative is a maximal $H \cong Fi_{23}$, and the associated permutation character is π. The subdegrees are $1, 31671, 275264$, and the collapsed adjacency matrices are

$$
\begin{pmatrix}
0 & 31671 & 0 \\
1 & 3510 & 28160 \\
0 & 3240 & 28431
\end{pmatrix}
$$

$$
\begin{pmatrix}
0 & 0 & 275264 \\
0 & 28160 & 247104 \\
1 & 28431 & 246832
\end{pmatrix}.
$$

The vertex-set of these orbital graphs may be taken to be the class of 3-transpositions of $Fi_{24} \cong G:2$. For the first orbital graph we join 3-transpositions d, e exactly when de has order 2 (equivalently $d \neq e$ and d, e commute), and for the second graph we join d, e exactly when de has order 3. These graphs were originally studied by B. Fischer [Fis69].

4.36.3 rank 4

G has no pseudo-permutation character of rank 4.

4.36.4 rank 5

G has no pseudo-permutation character of rank 5.

4.37 Automorphism group $Fi_{24} = Fi'_{24}:2$ of Fi'_{24}

A presentation for $3 \cdot Fi_{24}$ is given in the section for Fi'_{24}. In [HS95], a presentation is given for Fi_{24}, as a 3-transposition group generated by five 3-transpositions.

4.37.1 rank 2

G has no faithful pseudo-permutation character of rank 2.

4.37.2 rank 3

(i) $\pi = \chi_1 + \chi_5 + \chi_7$ of degree $306936 = 1 + 57477 + 249458$.

The restriction of π to Fi_{24}' is the permutation character in case (i) for Fi_{24}'. Thus $H \cong Fi_{23} \times 2$, and the collapsed adjacency matrices are the same as for that case.

4.37.3 rank 4

G has no pseudo-permutation character of rank 4.

4.37.4 rank 5

G has no pseudo-permutation character of rank 5.

4.38 The Fischer Baby Monster group B

In [Iva94], A.A. Ivanov derives a presentation for the Baby Monster group B, by proving that $Y_{433} \cong 2^2.B$. In this Y_{433}, there is a visible $2^3.{}^2E_6(2){:}2 \cong Y_{333}$. Details and definitions can be found in [Iva94], where Y_{pqr} is defined as a certain finitely presented group. (This definition is somewhat different from the definition of Y_{pqr} in [CNS88], which is different again from the definition in the ATLAS.) The representation of Y_{433} on the cosets of Y_{333} is of degree 13571955000, and is well out of the range of current coset enumeration technology.

4.38.1 rank 2

G has no pseudo-permutation character of rank 2.

4.38.2 rank 3

G has no pseudo-permutation character of rank 3.

4.38.3 rank 4

G has no pseudo-permutation character of rank 4.

4.38.4 rank 5

(i) $\pi = \chi_1 + \chi_3 + \chi_5 + \chi_{13} + \chi_{15}$ of degree $13571955000 = 1 + 96255 + 9458750 + 4275362520 + 9287037474$.

By [MM85, ILLSS95] there is a unique class of subgroups of G of index 13571955000, a representative is a maximal $H \cong 2^{\cdot 2}E_6(2):2$, and the associated permutation character is π. The subdegrees are

$$1, 3968055, 23113728, 2370830336, 11174042880.$$

Using the information in [Iva92] on the orbital digraph corresponding to the suborbit of length 3968055, D.V. Pasechnik calculated the collapsed adjacency matrix A for this graph. This collapsed adjacency matrix had originally been computed much earlier by B. Fischer (see [Hig76]). The calculation of a collapsed adjacency matrix for such a large representation is beyond the scope of the methods described in this book.

The matrix A has distinct eigenvalues, allowing the collapsed adjacency matrices for the other orbital digraphs of this multiplicity-free representation to be determined from A (this uses some basic theory of Bose-Mesner algebras; see [BCN89, pp. 43–46]). This was done by D.V. Pasechnik, using a *Maple* program he wrote for this purpose.

The collapsed adjacency matrices are

$$\begin{pmatrix} 0 & 3968055 & 0 & 0 & 0 \\ 1 & 46134 & 0 & 2097152 & 1824768 \\ 0 & 0 & 69615 & 0 & 3898440 \\ 0 & 3510 & 0 & 837135 & 3127410 \\ 0 & 648 & 8064 & 663552 & 3295791 \end{pmatrix}$$

$$\begin{pmatrix} 0 & 0 & 23113728 & 0 & 0 \\ 0 & 0 & 405504 & 0 & 22708224 \\ 1 & 69615 & 0 & 6336512 & 16707600 \\ 0 & 0 & 61776 & 3592512 & 19459440 \\ 0 & 8064 & 34560 & 4128768 & 18942336 \end{pmatrix}$$

$$\begin{pmatrix} 0 & 0 & 0 & 2370830336 & 0 \\ 0 & 2097152 & 0 & 500170752 & 1868562432 \\ 0 & 0 & 6336512 & 368492544 & 1996001280 \\ 1 & 837135 & 3592512 & 423236608 & 1943164080 \\ 0 & 663552 & 4128768 & 412286976 & 1953751040 \end{pmatrix}$$

$$\begin{pmatrix} 0 & 0 & 0 & 0 & 11174042880 \\ 0 & 1824768 & 22708224 & 1868562432 & 9280947456 \\ 0 & 3898440 & 16707600 & 1996001280 & 9157435560 \\ 0 & 3127410 & 19459440 & 1943164080 & 9208291950 \\ 1 & 3295791 & 18942336 & 1953751040 & 9198053712 \end{pmatrix}.$$

There are no non-complete distance-regular generalized orbital graphs.

4.39 The Fischer-Griess Monster group M

G has no faithful pseudo-permutation character of rank less than or equal to 5.

5

Summary of the Representations and Graphs

In this chapter we give tables summarizing the transitive permutation representations and distance-regular generalized orbital graphs we have studied and classified in this book. In particular, Tables 5.1, 5.2, 5.3, and 5.4 list the primitive representations of respective ranks 2, 3, 4, and 5 of the sporadic almost simple groups, while Tables 5.5, 5.6, and 5.7 list the imprimitive representations of these groups for respective ranks 3, 4, and 5. These tables also give the number of distance-regular generalized orbital graphs for each of these representations. Finally, Table 5.8 gives a summary of the graphs of diameters 3 and 4 having a distance-transitive action by a sporadic almost simple group.

For each table of permutation representations, the column labelled 'G' gives the group being represented, the 'case' column gives the roman-numbered case for G in Chapter 4, where the representation is described in detail, the column labelled 'H' gives the point stabilizer, and 'degree' the degree of the representation. Notation of the form $K[.2]$ for G and $L[.2]$ for H denotes two representations of the same rank and degree: one of $K.2$ on the cosets of $L.2$, and the restriction of this representation to K on the cosets of L. In this situation there will be two entries in the 'case' column, corresponding respectively to K and $K.2$. Given an entry for a group G with point stabilizer H, the notation H (2 classes) denotes that there are two G-classes of subgroups isomorphic to H, and that these classes are interchanged by an outer automorphism of G, giving two equivalent (but not permutationally isomorphic) representations for G.

A line in a table of representations gives information on the representation of a group G acting on the (right) cosets of a subgroup H (or each

129

of $G = K, K.2$ acting on the cosets of $H = L, L.2$, respectively). We put an m in the 'd.r.' (distance-regular) column of this line to denote that exactly m of the generalized orbital digraphs for the permutation representation of G on the cosets of H are distance-regular graphs. The complete graph is included in this count. We put an n in the 'd.t.' (distance-transitive) column to denote that G acts distance-transitively on exactly n of these distance-regular (generalized) orbital digraphs. All these distance-regular and distance-transitive graphs are described in Chapter 4 in the sections for the corresponding representations. The new distance-regular graphs of diameter 2 for $O'N$, Co_2 and Fi_{22} are of particular interest.

We thus have a complete classification of the distance-regular graphs on which a sporadic almost simple group acts vertex-transitively with rank at most 5, as well as a classification of the (distance-regular) graphs of diameter at most 4 on which a sporadic almost simple group acts distance-transitively. It is now known [ILLSS95] that all distance-regular graphs on which a sporadic almost simple group acts primitively and distance-transitively have diameter at most 4, and so these graphs appear in our classification.

In Table 5.8, we summarize those simple connected graphs Γ of diameters 3 and 4 having a distance-transitive action by a sporadic almost simple group G. A check mark is put in the column labelled 'prim' if and only if G acts primitively on the vertices of Γ, 'n' denotes the number of vertices of Γ, and 'diam' is the diameter of Γ.

Table 5.1. *The rank 2 representations*

G	case	H	degree	d.r.	d.t.
M_{11}	(i)	M_{10}	11	1	1
M_{11}	(ii)	$L_2(11)$	12	1	1
M_{12}	(i)	M_{11} (2 classes)	12	1	1
$M_{22}[.2]$	$(i), [i]$	$L_3(4)[.2]$	22	1	1
M_{23}	(i)	M_{22}	23	1	1
HS	(i)	$U_3(5).2$ (2 classes)	176	1	1
M_{24}	(i)	M_{23}	24	1	1
Co_3	(i)	$McL.2$	276	1	1

Table 5.2. *The rank 3 primitive representations*

G	case	H	degree	d.r.	d.t.
M_{11}	(iv)	$M_9.2$	55	3	2
M_{12}	(ii)	$M_{10}.2$ (2 classes)	66	3	2
$M_{22}[.2]$	$(iii),[ii]$	$2^4:A_6[.2]$	77	3	2
M_{22}	(iv)	A_7 (2 classes)	176	3	2
$J_2[.2]$	$(i),[i]$	$U_3(3)[.2]$	100	3	2
M_{23}	(ii)	$L_3(4).2$	253	3	2
M_{23}	(ii)	$2^4:A_7$	253	3	2
$HS[.2]$	$(ii),[i]$	$M_{22}[.2]$	100	3	2
M_{24}	(iii)	$M_{22}.2$	276	3	2
M_{24}	(iv)	$M_{12}.2$	1288	3	2
$McL[.2]$	$(i),[i]$	$U_4(3)[.2]$	275	3	2
Ru	(i)	$^2F_4(2)$	4060	3	2
$Suz[.2]$	$(i),[i]$	$G_2(4)[.2]$	1782	3	2
Co_2	(i)	$U_6(2).2$	2300	3	2
$Fi_{22}[.2]$	$(i),[i]$	$2\cdot U_6(2)[.2]$	3510	3	2
Fi_{22}	(ii)	$O_7(3)$ (2 classes)	14080	3	2
Fi_{23}	(i)	$2\cdot Fi_{22}$	31671	3	2
Fi_{23}	(ii)	$O_8^+(3).S_3$	137632	3	2
$Fi'_{24}[.2]$	$(i),[i]$	$Fi_{23}[\times 2]$	306936	3	2

Table 5.3. *The rank 4 primitive representations*

G	case	H	degree	d.r.	d.t.
M_{11}	(v)	S_5	66	3	0
$M_{12}.2$	(ii)	$L_2(11).2$	144	7	0
$M_{12}.2$	(iii)	$L_2(11).2$	144	3	0
$M_{22}[.2]$	$(vii),[v]$	$2^4:S_5[.2]$	231	5	0
$J_2[.2]$	$(ii),[iii]$	$3\cdot PGL_2(9)[.2]$	280	5	0
M_{23}	(iii)	A_8	506	2	1
M_{23}	(iv)	M_{11}	1288	3	0
M_{24}	(vi)	$2^4:A_8$	759	2	1
M_{24}	(vii)	$2^6:3\cdot S_6$	1771	1	0
McL	(ii)	M_{22} (2 classes)	2025	1	0
$He.2$	(i)	$S_4(4).4$	2058	1	0
$Fi_{22}[.2]$	$(iv),[ii]$	$O_8^+(2).S_3[\times 2]$	61776	5	0
Co_1	(i)	Co_2	98280	1	0

Table 5.4. *The rank 5 primitive representations*

G	case	H	degree	d.r.	d.t.
M_{12}	(vi)	$L_2(11)$	144	3	0
M_{12}	(vii)	$M_9.S_3$ (2 classes)	220	2	0
J_1	(iii)	$L_2(11)$	266	2	1
$M_{22}[.2]$	$(ix),[vi]$	$2^3:L_3(2)[\times 2]$	330	2	1
$M_{22}[.2]$	$(xi),[ix]$	$M_{10}[.2]$	616	1	0
$J_2.2$	(v)	$2^{1+4}_-.S_5$	315	2	1
$HS[.2]$	$(iv),[iii]$	$L_3(4).2[.2]$	1100	1	0
$HS[.2]$	$(v),[iv]$	$S_8[\times 2]$	1100	1	0
M_{24}	(x)	$L_3(4).S_3$	2024	2	0
M_{24}	$(xiii)$	$2^6:(L_3(2)\times S_3)$	3795	1	0
$McL[.2]$	$(iv),[ii]$	$U_3(5)[.2]$	7128	1	0
$McL[.2]$	$(v),[iii]$	$3^{1+4}_+:2.S_5[.2]$	15400	1	0
He	(i)	$S_4(4).2$	2058	1	0
$Suz[.2]$	$(iii),[ii]$	$3\cdot U_4(3).2[.2]$	22880	2	1
$O'N$	(i)	$L_3(7).2$ (2 classes)	122760	3	0
Co_3	(iii)	HS	11178	1	0
Co_2	(iii)	$2^{10}:M_{22}.2$	46575	1	0
Co_2	(iv)	$2^{1+8}_+:S_6(2)$	56925	3	0
Ly	(i)	$G_2(5)$	8835156	1	0
Ly	(ii)	$3\cdot McL.2$	9606125	1	0
Co_1	(ii)	$3\cdot Suz.2$	1545600	1	0
B	(i)	$2\cdot{}^2E_6(2).2$	13571955000	1	0

Table 5.5. *The rank 3 imprimitive representations*

G	case	H	degree	d.r.	d.t.
M_{11}	(iii)	A_6	22	2	1
$M_{12}.2$	(i)	M_{11}	24	2	1

Table 5.6. *The rank 4 imprimitive representations*

G	case	H	degree	d.r.	d.t.
M_{11}	(vi)	$3^2:8$	110	2	0
M_{12}	(iii)	$PGL_2(9)$ (2 classes)	132	2	0
$M_{22}.2$	(iii)	$L_3(4)$	44	4	1
HS	(iii)	$U_3(5)$ (2 classes)	352	4	2
$HS.2$	(ii)	$U_3(5).2$	352	4	2
Co_3	(ii)	McL	552	4	2

Table 5.7. *The rank 5 imprimitive representations*

G	case	H	degree	d.r.	d.t.
M_{12}	(iv)	S_6 (2 classes)	132	2	0
M_{12}	(v)	$L_2(11)$	144	13	0
$M_{12}.2$	(iv)	$M_{10}.2$	132	2	0
$M_{22}[.2]$	$(x), [viii]$	$2^4 : L_2(5)[.2]$	462	2	0
M_{24}	(xi)	M_{12}	2576	2	0
M_{24}	(xii)	$2^6 : 3 \cdot A_6$	3542	2	0
Ru	(ii)	$^2F_4(2)'$	8120	2	0
Co_2	(ii)	$U_6(2)$	4600	2	0
Fi_{23}	(iii)	$O_8^+(3).3$	275264	2	0
Fi_{23}	(iv)	$O_8^+(3).2$	412896	2	0

Table 5.8. *Distance-transitive graphs of diameters 3 and 4*

G	prim	n	diam	intersection array
M_{23}	✓	506	3	$\{15, 14, 12; 1, 1, 9\}$
M_{24}	✓	759	3	$\{30, 28, 24; 1, 3, 15\}$
J_1	✓	266	4	$\{11, 10, 6, 1; 1, 1, 5, 11\}$
$M_{22}[.2]$	✓	330	4	$\{7, 6, 4, 4, ; 1, 1, 1, 6\}$
$J_2.2$	✓	315	4	$\{10, 8, 8, 2; 1, 1, 4, 5\}$
$Suz[.2]$	✓	22880	4	$\{280, 243, 144, 10; 1, 8, 90, 280\}$
$M_{22}.2$		44	3	$\{21, 20, 1; 1, 20, 21\}$
HS		352	3	$\{175, 72, 1; 1, 72, 175\}$
HS		352	3	$\{175, 102, 1; 1, 102, 175\}$
$HS.2$		352	3	$\{50, 49, 36; 1, 14, 50\}$
$HS.2$		352	3	$\{126, 125, 36; 1, 90, 126\}$
Co_3		552	3	$\{275, 112, 1; 1, 112, 275\}$
Co_3		552	3	$\{275, 162, 1; 1, 162, 275\}$

Bibliography

[Asc94] M. Aschbacher, *Sporadic Groups*, Cambridge University Press, 1994.

[AS85] M. Aschbacher and L. Scott, Maximal subgroups of finite groups, *J. Algebra* **92** (1985), 44–80.

[Ban72] E. Bannai, Maximal subgroups of low rank of finite symmetric and alternating groups, *J. Fac. Sci. Univ. Tokyo* **18** (1972), 475–486.

[BI84] E. Bannai and T. Ito, *Algebraic Combinatorics I: Association Schemes*, Benjamin, New York, 1984.

[BL96] T. Breuer and K. Lux, The multiplicity-free permutation characters of the sporadic simple groups and their automorphism groups, *Comm. Algebra* **24** (1996), 2293–2316.

[BCN89] A.E. Brouwer, A.M. Cohen and A. Neumaier, *Distance-Regular Graphs*, Springer, Berlin and New York, 1989.

[BDD88] F. Buekenhout, A. Delandtsheer and J. Doyen, Finite linear spaces with flag-transitive automorphism groups, *J. Combin. Theory, Series A* **49** (1988), 268–293.

[BDDKLS90] F. Buekenhout, A. Delandtsheer, J. Doyen, P.B. Kleidman, M.W. Liebeck and J. Saxl, Linear spaces with flag-transitive automorphism groups, *Geom. Ded.* **36** (1990), 89–94.

[Bur11] W. Burnside, *Theory of groups of finite order*, Cambridge University Press, 1911 (2nd Edn), reprinted by Dover, New York, 1955.

[But91] G. Butler, *Fundamental Algorithms for Permutation Groups*, Lecture Notes in Computer Science **559**, Springer, Berlin and New York, 1991.

[Cam81] P.J. Cameron, Finite permutation groups and finite simple groups, *Bull. London Math. Soc.* **13** (1981), 1–22.

[CvL91] P.J. Cameron and J.H. van Lint, *Designs, Graphs, Codes and their Links*, LMS Student Texts **22**, Cambridge University Press, 1991.

[Can84] J.J. Cannon, An introduction to the group theory language, Cayley, in *Computational Group Theory* (M.D. Atkinson, ed.), Academic Press, New York and London, 1984.

[CP95] J. Cannon and C. Playoust, *An Introduction to MAGMA*, School of Mathematics and Statistics, University of Sydney, 1995.

[CGGLMW92] B.W. Char, K.O. Geddes, G.H. Gonnet, B.L. Leong, M.B. Monagan and S.M. Watt, *First Leaves: A Tutorial Introduction to Maple V*, Springer, Berlin and New York, 1992.

[CCNPW85] J.H. Conway, R.T. Curtis, S.P. Norton, R.A. Parker and R.A. Wilson, *An ATLAS of Finite Groups*, Clarendon Press, Oxford, 1985.

[CNS88] J.H. Conway, S.P. Norton, and L.H. Soicher, The Bimonster, the group Y_{555}, and the projective plane of order 3, in *Computers in Algebra* (M.C. Tangora, ed.), Marcel Dekker, New York, 1988, pp. 27–50.

[CP92] J.H. Conway and A.D. Pritchard, Hyperbolic reflections for the Bimonster and $3Fi_{24}$, in *Groups, Combinatorics and Geometry* (M.W. Liebeck and J. Saxl, eds.), L.M.S. Lecture Notes **165**, Cambridge University Press, 1992, pp. 24–45.

[CS88] J.H. Conway and N.J.A. Sloane, *Sphere Packings, Lattices and Groups*, Springer, Berlin and New York, 1988.

[CFYT94] G. Cooperman, L. Finkelstein, B. York and M. Tselman, Constructing permutation representations for large matrix groups, in *Proc. ISSAC '94*, ACM, 1994, pp. 134–138.

[CKS76] C.W. Curtis, W.M. Kantor and G.M. Seitz, The 2-transitive permutation representations of the finite Chevalley groups, *Trans. Amer. Math. Soc.* **218** (1976), 1–57.

[Cuy89] H. Cuypers, Low rank permutation representations of the finite groups of Lie type, Part 1 of: Geometries and permutation groups of small rank, Doctoral thesis, Rijksuniversiteit, Utrecht, 1989.

[FIK90] I.A. Faradžev, A.A. Ivanov and M.H. Klin, Galois correspondence between permutation groups and cellular rings (association schemes), *Graphs and Combinatorics* **6** (1990), 303–332.

[FK91] I.A. Faradžev and M.H. Klin, Computer package for computations with coherent configurations, in *Proc. ISSAC '91*, ACM, 1991, pp. 219–223.

[FKM94] I.A. Faradžev, M.H. Klin and M.E. Muzichuk, Cellular rings and groups of automorphisms of graphs, in *Investigations in Algebraic Theory of Combinatorial Objects* (I.A. Faradžev, A.A. Ivanov, M.H. Klin and A.J. Woldar, eds.), Kluwer Academic Publishers, 1994, pp. 1–153.

[FT63] W. Feit and J.G. Thompson, Solvability of groups of odd order, *Pacific J. Math.* **13** (1963), 775–1027.

[Fis69] B. Fischer, Finite groups generated by 3-transpositions, University of Warwick Lecture Notes, 1969.

[Fou69] D.A. Foulser, Solvable primitive permutation groups of low rank, *Trans. Amer. Math. Soc.* **143** (1969), 1–54.

[Gol95] H.W. Gollan, A new existence proof for Ly, the sporadic simple group of R. Lyons, Habilitationsschrift, Preprint 30, Institute for Experimental Mathematics, University of Essen, 1995.

[Gor68] D. Gorenstein, *Finite Groups*, Harper and Row, New York, 1968.

[Gor82] D. Gorenstein, *Finite Simple Groups*, Plenum Press, New York and London, 1982.

[HS95] J.I. Hall and L.H. Soicher, Presentations of some 3-transposition groups, *Comm. Algebra* **23** (1995), 2517–2559.

[Hal76] M. Hall, Jr, Group properties of Hadamard matrices, *J. Austral. Math. Soc. (A)* **21** (1976), 247–256.

[Her74] C. Hering, Transitive linear groups, and linear groups which contain irreducible subgroups of prime order, *Geom. Dedicata* **2** (1974), 425–460.

[Her85] C. Hering, Transitive linear groups, and linear groups which contain irreducible subgroups of prime order II, *J. Algebra* **93** (1985), 151–164.

[Hig67] D.G. Higman, Intersection matrices for finite permutation groups, *J. Algebra* **6** (1967), 22–42.

[Hig75] D.G. Higman, Coherent configurations, Part I: Ordinary representation theory, *Geom. Dedicata* **4** (1975), 1–32.

[Hig76] D.G. Higman, Coherent configurations, Part II: Weights, *Geom. Dedicata* **5** (1976), 413–424.

[Hig76] D.G. Higman, A monomial character of Fischer's Baby Monster, in *Proc. of the Conference on Finite Groups* (W.R. Scott and F. Gross, eds.), Academic Press, New York and London, 1976, pp. 277–283.

[HS68] D. Higman and C. Sims, A simple group of order 44,352,000, *Math. Z.* **105** (1968), 110–113.

[Hig69] G. Higman, On the simple group of D.G. Higman and C.C. Sims, *Illinois J. Math.* **13** (1969), 74–80.

[Hup57] B. Huppert, Zweifach transitive, auflösbare Permutationsgruppen, *Math. Z.* **68** (1957), 126–150.

[Isa76] I.M. Isaacs, *Character Theory of Finite Groups*, Academic Press, New York and London, 1976.

[Iva92] A.A. Ivanov, A geometric characterization of Fischer's baby monster, *J. Algebraic Combinatorics* **1** (1992), 45–69.

[Iva94] A.A. Ivanov, Presenting the baby monster, *J. Algebra* **163** (1994), 88–108.

[IKF82] A.A. Ivanov, M.H. Klin and I.A. Faradžev, Primitive representations of nonabelian simple groups of order less than 10^6, Part 1, VNIISI Preprint, Moscow, 1982. [In Russian]

[IKF84] A.A. Ivanov, M.H. Klin and I.A. Faradžev, Primitive representations of nonabelian simple groups of order less than 10^6, Part 2, VNIISI Preprint, Moscow, 1984. [In Russian]

[ILLSS95] A.A. Ivanov, S.A. Linton, K. Lux J. Saxl and L.H. Soicher, Distance-transitive representations of the sporadic groups, *Comm. Algebra* **23** (1995), 3379–3427.

[JLPW95] C. Jansen, K. Lux, R. Parker and R. Wilson, *An Atlas of Brauer Characters*, Clarendon Press, Oxford, 1995.

[JW96] C. Jansen and R.A. Wilson, The minimal faithful 3-modular representation for the Lyons group, *Comm. Algebra* **24** (1996), 873–879.

[Kan81] W.M Kantor, Some geometries that are almost buildings, *Europ. J. Combinatorics* **2** (1981), 239–247.

[Kan85] W.M. Kantor, Homogeneous designs and geometric lattices, *J. Combin. Theory (A)* **38** (1985), 66–74.

[KL82] W.M. Kantor and R.A. Liebler, The rank 3 permutation representations of the finite classical groups, *Trans. Amer. Math. Soc.* **71** (1982), 1–71.

[KPW89] P.B. Kleidman, R.A. Parker, and R.A. Wilson, The maximal subgroups of the Fischer group Fi_{23}, *J. London Math. Soc.* (2) **39** (1989), 89–101.

[Leo80] J.S. Leon, On an algorithm for finding a base and strong generating set for a group given by generating permutations, *Math. Comp.* **35** (1980), 941–974.

[Lie87] M.W. Liebeck, The affine permutation groups of rank 3, *Proc. London Math. Soc. (3)* **54** (1987), 477–516.

[LPS88] M.W. Liebeck, C.E. Praeger and J. Saxl, On the O'Nan-Scott theorem for finite primitive permutation groups, *J. Austral. Math. Soc. (A)* **44** (1988), 389–396.

[LPS90] M.W. Liebeck, C.E. Praeger and J. Saxl, *The maximal factorizations of the finite simple groups and their automorphism groups*, Memoirs of the AMS **432**, AMS, Providence, 1990.

[LS86] M.W. Liebeck and J. Saxl, The finite primitive permutation groups of rank 3, *Bull. London Math. Soc.* **18** (1986), 165–172.

[LLS95] S.A. Linton, K. Lux and L.H. Soicher, The primitive distance-transitive representations of the Fischer groups, *Experimental Math.* **4** (1995), 235–253.

[LW91] S.A. Linton and R.A. Wilson, The maximal subgroups of the Fischer groups Fi_{24} and Fi'_{24}, *Proc. London Math. Soc. (3)* **63** (1991), 113–164.

[Mai1895] E. Maillet, Sur les isomorphes holoédriques et transitifs des groupes symetriques ou alternés, *J. Math. Pures Appl. (5)* **1** (1895), 5–34.

[MM85] V.D. Mazurov and N.P. Mazurova, Large subgroups of the simple group F_2, *Mat. Zametki* **37** (1985), 145–151. [In Russian]

[McK90] B.D. McKay, *nauty* user's guide (version 1.5), Technical report TR-CS-90-02, Computer Science Department, Australian National University, 1990.

[McL69] J. McLaughlin, A simple group of order 898,128,000, in *Symposium on Finite Groups* (R. Brauer and C. Sah, eds.), Benjamin, New York, 1969, pp. 109–111.

[MNP85] W. Meyer, W. Neutsch and R. Parker, The minimal 5-representation of Lyon's sporadic group, *Math. Ann.* **272** (1985), 29–39.

[Neu82] J. Neubüser, An elementary introduction to coset table methods in computational group theory, in *Groups – St. Andrews 1981* (C.M. Campbell and E.F. Robertson, eds.), LMS Lecture Notes **71**, Cambridge University Press, 1982, pp. 1–45.

[NST94] P.M. Neumann, G.A. Stoy and E.C. Thompson, *Groups and Geometry*, Oxford University Press, 1994.

[PSY87] C.E. Praeger, J. Saxl and K. Yokoyama, Distance transitive graphs and finite simple groups, *Proc. London Math. Soc. (3)* **55** (1987), 1–21.

[Sch95] M. Schönert, et al., GAP: Groups, Algorithms and Programming, Lehrstuhl D für Mathematik, RWTH Aachen, 1995.

[Sco80] L.L. Scott, Representations in characteristic p, in *The Santa Cruz Conference on Finite Groups, Proceedings of Symposia in Pure Mathematics* **37**, AMS, 1980, pp. 318–331.

[Sei74] G.M. Seitz, Small rank permutation representations of finite Chevalley groups, *J. Algebra* **28** (1974), 508–517.

[Sim67] C.C. Sims, Graphs and finite permutation groups, *Math. Z.* **95** (1967), 76–86.

[Sim73] C.C. Sims, The existence and uniqueness of Lyons' group, in *Finite Groups '72* (T. Gagen, M.P. Hale, Jr and E.E. Shult, eds.), North-Holland, Amsterdam, 1973, pp. 138–141.

[Soi85] L.H. Soicher, Presentations of some finite groups, PhD thesis, Cambridge, 1985.

[Soi87a] L.H. Soicher, Presentations for Conway's group Co_1, *Math. Proc. Camb. Phil. Soc.* **102** (1987), 1–3.

[Soi87b] L.H. Soicher, Presentations of some finite groups with applications to the O'Nan simple group, *J. Algebra* **108** (1987), 310–316.

[Soi88] L.H. Soicher, Presentations for some groups related to Co_1, in *Computers in Algebra* (M.C. Tangora, ed.), Marcel Dekker, New York, 1988, pp. 151–154.

[Soi90] L.H. Soicher, A new existence and uniqueness proof for the O'Nan group, *Bull. London Math. Soc.* **22** (1990), 148–152.

[Soi91] L.H. Soicher, A new uniqueness proof for the Held group, *Bull. London Math. Soc.* **23** (1991), 235–238.

[Soi93a] L.H. Soicher, The Lyons group has no distance-transitive representation, in *Finite Geometry and Combinatorics* (F. De Clerck et al., eds.), LMS Lecture Notes **191**, Cambridge University Press, 1993, pp. 355–358.

[Soi93b] L.H. Soicher, GRAPE: a system for computing with graphs and groups, in *Groups and Computation* (L. Finkelstein and W.M. Kantor, eds.), DIMACS Series in Discrete Mathematics and Theoretical Computer Science **11**, AMS, 1993, pp. 287–291.

[Soi95] L.H. Soicher, Yet another distance-regular graph related to a Golay code, *Electronic J. Combinatorics* **2** (1995), #N1.

[Suz69] M. Suzuki, A simple group of order 448,345,497,600, in *Symposium on Finite Groups* (R. Brauer and C. Sah, eds.), Benjamin, New York, 1969, pp. 113–119.

[Wei91] R.M. Weiss, A geometric characterization of the groups M_{12}, He and Ru, *J. Math. Soc. Japan* **43** (1991), 795–814.

[Wil86] R.A. Wilson, Maximal subgroups of sporadic groups, in *Groups – St. Andrews 1985* (C.M. Campbell and E.F. Robertson, eds.), LMS Lecture Notes **121**, Cambridge University Press, 1986, pp. 352–358.

[Wil] R.A. Wilson, An ATLAS of sporadic group representations, in *The ATLAS 10 Years On* (R.T. Curtis and R.A. Wilson, eds.), Cambridge University Press, to appear.

Index

139